Programming of Life

Programming of Life

Donald E. Johnson

Published by Big Mac Publishers
Printed and bound in the United States of America

Author: Donald E. Johnson

Cover photographs ©iStockphoto.com/David Marchal
©iStockphoto.com/loops7
Cover Illustration / Design / Cartoon: Jess Nilo Design + Illustration

Library of Congress Control Number: 2010933477
Library of Congress subject headings:
QH325 Life--Origin
QH359 Evolution (Biology)
QH371.5 Macroevolution
Z665 Information science

BIASC / BASIC Classification Suggestions:
1. sh85014173 Biological control systems
2. sh2006008029 Biosemiotics
3. sh00003585 Bioinformatics
4. sh85133362 Teleology
5. sh85066291 Information theory in biology

ISBN-13: 978-0-9823554-6-6
ISBN-10: 0-9823554-6-7
1.1

Published by Big Mac Publishers
www.bigmacpublishers.com /Sylacauga, Alabama
Printed and bound in the United States of America

A Quick Guide to Reading Each Chapter (by chapter number)

1. This is a foundational chapter for appreciating the scientific-notation numbers used throughout the book. The fundamentals of probability (possible, impossible, probable, feasible) must also be understood. This will be easy reading if you're strong in math, but may take an extraordinary amount of time if you've found math to be challenging, and if you want to totally understand the concepts. If all you want is a "feel" for the numbers, the time will be considerably less (and you can always come back to this chapter if deeper understanding is desired later).
2. Understanding the types of information and data, and information's expression and communication are critical.
3. The development of computer hardware and software over time gives insight into the processes required as complexity increases.
4. This chapter should give an appreciation for the cybernetic complexities of even "simple" life, but needn't be thoroughly understood since the purpose isn't to make you a biologist. The terms that will be used in the following chapters are **highlighted** so that particular attention may be given to them. A critique of proposed origin-of-life scenarios should provide information that is usually missing when those scenarios are presented.
5. Shannon information theory is applied to the information storage and communication structures of life. Shannon channel capacity rules out "simpler" information structures in life.
6. The algorithmic prescriptive information in life is evaluated.
7. Complex Shannon, prescriptive, and functional information and associated information processing systems within life are highlighted and ramifications evaluated.
8. The mechanisms of biological evolution are evaluated using information sciences. This chapter is particularly important for evaluating the plausibility of evolutionary scenarios.
9. What are the impacts of life's information? Is physicality a sufficient explanation? Should other avenues be investigated?

Table of Contents

A Quick Guide to Reading Each Chapter iv

Introduction vii

1 Math Basics: Probability and Large or Small Numbers 1

2 Information Basics: Data and Information Types 7

3 Evolution of Computer Hardware and Software 13

4 Life Basics 17

5 Shannon Information in Life 33

6 Prescriptive Programming Information in Life 39

7 Combining Life's Information Types 47

8 Programming Increasing Complexity in Life 55

9 Unresolved Difficulties of Life's Information Requirements 79

References 85

Appendices

A Logarithms, Probability, and Other Math 103

B Comparison of Computer Disk Drive and DNA 104

C Life Details and Origin Speculations 106

D Shannon Information Technical Details 112

E Functional Information Technical Examples 113

F What Happened to Darwinism? 116

G Scientific Falsification and Specific Hypotheses 119

H Philosophical Hindrances to Scientific Truth 121

I Index of Definitions/Descriptions 127

Acknowledgments

The author wishes to thank Richard Hughes for his encouragement.

The author thanks the peer professionals (including those who wish to remain anonymous – see Chapter 9) who reviewed this book and offered invaluable suggestions for improvements and corrections of errors. Those professionals include:

David L. Abel, Director, The Gene Emergence Project
Robert Sheldon, NASA-funded Physicist
Josh Mitchell, Biologist / Regulatory Project Manager
Jonathan Bartlett, Director of Technology / Author / Speaker

Thanks also to Greg McElveen and Helen Fox for guidance in preparing the manuscript and to Jim Pappas for helpful suggestions.

Comments or suggestions for error corrections or improvements sent to don@scienceintegrity.net will be greatly appreciated.

Errata and updates are posted at scienceintegrity.net

See scienceintegrity.net for information on bulk purchases and a companion DVD.

Introduction

This book is an expansion of the information of life topics introduced in *Probability's Nature and Nature's Probability: A Call to Scientific Integrity* [Joh09I], which was written for scientists, and the "Lite" version [Joh09L], which was written for non-scientists. This book assumes the reader is not a scientist, so many technical details (including information theory's equations and several of life's structures) are included in the appendices so that the reader needn't become overwhelmed by the math and other technical language during a normal reading. Appendices also have supplemental material that, while important, isn't critical for understanding the book's main thrust. There are some technical details that must be presented during a normal reading, but a thorough understanding is usually not needed in order to appreciate the cybernetic complexities involved. The final page has an index of frequently-used terms. You may find this invaluable when encountering a familiar term whose meaning you want to verify.

This book presents the basis for bioinformatics, which is the study of the information in life. Bioinformatics includes a very broad range of scientific disciplines, but this book introduces the basis on which all the studies are built. By the time you finish this book, you will have a fairly good understanding of basic bioinformatics, as well as two important subsets, biosemiotics and biocybernetics.

Semiotics deals with symbols (signs) and their meanings. For example, the word "dog" is more than blotches of ink on a page, but is a symbol made up of other symbols (letters) to represent a particular type of animal that may be a pet. Biosemiotics is the characterization of the symbolic representations within life, which is filled with digitally-coded symbolic messages. Biocybernetics involves self-sustaining systems that integrate different levels of information and its processing, including controls and feedback, within biological systems. This is where "programming of life" fits in, as life's information processing systems involve thousands (or millions) of computer programs executing on thousands of interacting computers within each cell of an organism (an adult human has about 300 trillion cells). Don't be concerned if you don't understand those three bio-words at this point, as you haven't even reached chapter one yet.

This book draws attention to known facts that are usually overlooked or down-played when scenarios for the origin of life or Darwinian evolution are presented. The scenarios presented so far have not adequately addressed the complex functional information of life, especially the fact that life contains a multitude of complex programming algorithms whose origin by physical interactions cannot be explained using information science. As an information scientist, the author believes that the time has come to seriously look at the facts and consider different avenues of investigation that may provide theories that are scientifically testable. At the International Conference on Bioinformatics he asked, *"Does Bioinformatics support traditional biological views, or will it point in new directions, perhaps proposing other mechanisms for possible testing?"* [Joh04] Note: all quotes are italicized so that the reader can quickly identify them as quotes, identified by author and year.

While acknowledging that science continues to gain new insights (as should be the case), the claim that "we don't have a natural explanation yet, but we will someday" is not a scientific statement. It amounts to a "naturalism of the gaps" dogma. When that dogma violates known science, particularly information science, perhaps it's time to reevaluate stances that purport to be science, but are actually pseudo-scientific speculations.

This book will deal only with real science when claiming something as a fact. Any particular philosophical or theological view will not be endorsed as that would fall outside scientific investigation.

There are many more references than should be necessary, but they are included to document, typically in the scientists' own words, their views on items which may not be widely acknowledged. Most of the quotes are from scientists who believe that physicality is the only valid science, so don't take their statements to imply their support of anything else (even if the statements provide food for thought as to the adequacy of physical naturalism). It is important to realize that, contrary to a widespread belief of mainline science, the lack of scientific basis for many scenarios purported to be "true" is acknowledged by many scientists.

1 Math Basics: Probability and Large or Small Numbers

This short chapter may be skimmed by those already familiar with the topics, but to understand what the numbers mean in the rest of this book, it is vital to understand the concepts presented in this chapter. Information often involves probability, which involves the expression of large or small numbers. If the concepts are unfamiliar, it may require proceeding very slowly, until each item is understood. This chapter will clarify the scientific meaning of terms like "possible," "impossible," and "probable." For example, if a weather forecaster states "it will probably rain tomorrow," and it doesn't rain, that doesn't make the prediction wrong (unless the probability of rain were actually less than 50%). On the other hand, the prediction of 100% chance of rain tomorrow is probably wrong even if it does rain, since such a prediction probably cannot be made with absolute (100%) certainty (note the use of "probably" in this sentence since it cannot be proved with 100% certainty that one cannot predict the weather with 100% certainty).

Scientific or exponential notation is convenient when expressing very large or very small numbers. The Richter scale for earthquake magnitude is based exponentially, so that a magnitude 5 earthquake is 100 times (2 orders of magnitude) as strong as a magnitude 3 and 100 times weaker than a magnitude 7 quake. Since 6 is 20% higher than 5, if one is thinking linearly, rather than exponentially, one may visualize a magnitude 6 quake as only 20% stronger than a magnitude 5 quake, instead of the 900% stronger (10 times) that it really is. Exponential examples (take note of the exponent) include:

4,000,000,000 = 4 billion = 4×10^9

(count digits to the right of the first digit)

0.000001 = 1 millionth = $1/10^6 = 10^{-6}$

(count to the right of the "." including the first non-zero digit)

If all you need is a "feel" for the numbers, use the exponent as the number of zeros following or preceding (if negative) the number.

A 1 carat diamond contains approximately 10^{22} carbon atoms. A diamond containing 10 trillion (10^{13}) atoms would be a billionth (10^{-9}) carat diamond. A man hoping to impress his fiancé with a diamond containing a billion trillion (10^{21}) atoms, may find her unimpressed with the tenth carat ring. At 4800 characters per page, 10^{21} characters would require a stack of pages approximately 130 times the distance to the sun.

Notice that each unit change in exponent is a factor of ten more or less (all of the following have the same value).

$10^9 = 10 \times 10^8 = 10^{10}/10 = 1000 \times 10^6 = 1/10^{-9} = 10/10^{-8}$

For fun, use scientific prefixes for pico (10^{-12}), micro (10^{-6}), milli (10^{-3}), centi (10^{-2}), mega (10^6), giga (10^9), and tera (10^{12}), to verify: 10^{12} microphones = 1 megaphone, 10^{-1} centipede = 1 millipede, 10^6 bicycles = 2 megacycles, 10^{-6} fish = 1 microfiche, 10^{13} pins = 10 terrapins, 10^{-24} teraboo = 1 picoboo, and 10^{21} piccolos = 1 gigolo.

Numbers expressed exponentially may appear to be considerably different than what they represent. For example, a googol (not to be confused with the Google search engine) is 10^{100}, but is physically a totally hypothetical number since the maximum estimate of number of atoms in the Universe is 10^{80} [Sag79] (most estimates are many times less than this maximum, typically 10^{76} - 10^{78}). There is not a googol of anything physical (except maybe light photons) in the known Universe.

Logarithms can be used to calculate the exponent for any base, but are beyond the math requirements for this book. Logarithms are used for calculating many of the numbers presented in this book, but the details of those calculations are not usually presented, except in the appendices.

The law of probability expresses the likelihood of a particular outcome from within the set of possible outcomes. Probability has a range of 0 (impossible) to 1 (certain). Note that percent may be used to represent probability by using a range from 0% to 100%. For example, 50% is the same as 0.5, and 100% probability is a certainty (1).

Rolling a die has a probability of 1/6 (i. e. 1-in-6) for any particular number. Rolling a five 10 times in a row results in a 1/6 probability of a five on the next die roll since chance has no causative effect. Chance expresses likelihood, not cause – nothing is "caused by chance." Some people who don't understand chance play the lottery by betting on their "lucky" numbers, betting on the numbers that have been chosen most frequently, or betting on the numbers that have been chosen least frequently (reasoning that those numbers would have to be chosen more frequently in order to catch-up with their a priori probability).

For example, in the California SuperLotto lottery, one picks five different numbers (in any order) from 1 to 47 and one MEGA number from 1 to 27. If those numbers match the numbers drawn by the lottery, a win results (this discussion will be limited to a jackpot win, not a partial win by matching some numbers). The order of correct numbers chosen makes no difference (any incorrect number makes winning impossible). The first choice has a probability of 5/47 of being correct since there are

5 correct numbers in the 47 possibilities. If the first number is correct, the second number chosen has a probability of 4/46 since there are only 4 correct numbers out of the 46 numbers remaining. Similarity, the third through fifth choices have probabilities of being correct of 3/45, 2/44, and 1/43, respectively. The final "MEGA" number has a probability of 1/27 since the possibilities are 1 to 27.

The probability of independent consecutive events is the product of the probabilities of the individual events. The law of probability calculates that any arbitrary choice of numbers (regardless of how chosen) has a probability of winning the jackpot of one in 41,416,353, with the product of all six correct probabilities being 2.4×10^{-8} (= 1/41,416,353). Since the probability of winning plus the probability of not winning is 1 (a certainty), the probability of failing to win (P_f) the jackpot is 0.999999976 ($= 1 - 2.4 \times 10^{-8}$), a figure you won't see in lottery promotions. The probability of n consecutive failures is P_f^n (P_f times itself n times). To compute the number of trials needed to make a win as likely as continued consecutive failures, wave the logarithm magic wand (Appendix A) over the equation $0.5 = P_f^n$ to get $n = 2.9 \times 10^7$. A twice-a-week lottery player becomes a probable winner only after 2,767 centuries.

If a winner buys a single ticket for the next lottery, the chance of winning the second time is 2.4×10^{-8}. Note, however, that the probability of winning twice in a row is $(2.4 \times 10^{-8})^2$, or 5.8×10^{-16}. This is obviously very unlikely to happen. A first win results in people attributing the win to chance luck, whereas additional wins result in people attributing the wins to a rigged contest because of the extreme unlikelihood of such an outcome by chance.

The probability of selecting a particular atom of the Universe randomly is 10^{-80}. The probability of rolling ten fives in a row is $(1/6)^{10} = 1/60466176 = 1.6 \times 10^{-8}$. The probability of tossing 20 consecutive heads is 0.5^{20} or 9.53×10^{-7}. The probability of being dealt a royal flush from a deck of cards is:

$$(20/52) \times (4/51) \times (3/50) \times (2/49) \times (1/48) = 1/649740 = 1.54 \times 10^{-6}.$$

Assume the 48 contiguous US states were covered with densely-packed [Ste99] trees, each with 200,000 leaves [Wi-Web]. If one leaf were marked, and all leaves mixed together, the probability of choosing the marked leaf would be 3.1×10^{17}. If the 48 states were covered 18 inches deep in cents and a single 1943 copper cent were tossed into the pile, the probability of randomly choosing the 1943 coin would be 10^{-17}(about 58 times less likely than two consecutive lottery wins).

Note that increasing the number of attempts for something that is impossible will not increase its likelihood since the probability of failure is $(1 - 0)^n = 1$ for any n, so the probability of success is still 0. Likewise, the probability of a certainty doesn't decrease by having more trials since the probability of failure is $(1 - 1)^n = 0$ and 1 - 0 is still 1 (certain). If the probability isn't 0, theoretically it could happen. For example, the probability of throwing 300 consecutive heads is 0.5^{300} or 4.91×10^{-91}. Since chance has no causative effect, the desired pattern could happen on the first attempt, but it is less likely than 12 consecutive lottery wins.

A possible outcome becomes probable when its probability is at least 0.5 since any lower probability makes it more likely not to happen. For example, in rolling a pair of dice, it is probable that the sum will be greater than 6 since there are 21 of the 36 possible combinations that add to greater than 6, so that the probability is 21/36 (= 0.583) of obtaining a sum higher than 6.

Sometimes the probability of an outcome is uncertain. For example, if a coin being tossed may bounce differently when landing heads or tails, each may have a probability different than 0.5. Is it possible (non-zero probability) for a tossed die to end up on-edge? Just because an outcome hasn't yet been observed doesn't necessarily mean it couldn't happen. Perhaps a detailed analysis of cube motions could determine the on-edge probability. If it wasn't impossible, then it may be desirable to include it in the probability expression, although it seems insignificant for any practical use for die tosses. In the context of science, "possible" means a non-zero probability, which should only be used when known science demonstrates that to be true. It would not be scientifically accurate to state, "it is possible that the die may end up on-edge, reducing the 1/6 probability of each numeric result" unless one first demonstrates with known science that the on-edge result is possible. In case you're wondering, there have been reported instances of coins ending up on-edge [Mur93], and also of coin tosses having biased probability [Dia07].

If a die is rolled 100 times with each roll recorded, a very improbable pattern of digits will result, with $P = 1.5 \times 10^{-78}$ ($= 6^{-100}$) probability for that pattern. This pattern is extremely improbable, but if any outcome was acceptable, each roll had a probability of 1 of being correct. As has been shown, increasing the number of trials for a certainty does not reduce its probability, so $P = 1^{100} = 1$. The number of trials (n) to make a repeat of the pattern probable may be found by waving Taylor's magic wand (see Appendix A) over the math expression to yield $n = 4.6 \times 10^{79}$.

If a roll can be done in one second, the first improbable pattern occurs after only 100 seconds. The repeat of that pattern becomes probable only after 1.4×10^{70} centuries, which is approximately 10^{62} times longer than the oldest estimate of the Universe's age.

There are computer solutions that "explode" with increasing problem size. An example is the traditional solution to the Traveling Salesman Problem (TSP), in which the minimum traversal path is found to visit each city exactly once. The run-time of the traditional TSP algorithm (step-by-step solution) for n cities is proportional to the mathematical product of n times the run-time for one less city. This makes using that algorithm infeasible for more than a few cities. The time (t_n) for n cities has the relationship $t_n = nt_{n-1}$, so that $t_4 = 4t_3$, $t_5 = 5t_4$, $t_6 = 6t_5$, etc. If the solution for 10 cities takes 10 seconds, a solution for 11 cities would take 110 seconds, a solution for 15 cities would take over 1000 hours, and a 20-city solution would take 2,124 centuries, which is clearly infeasible since the salesman would die long before then. Dictionaries (e.g. – Random House©, Inc. 2006 and The American Heritage® Dictionary of the English Language, Fourth Edition) give definitions of infeasible as impracticable or unworkable, *"not capable of being carried out or put into practice."*

Some may reject the term "infeasible" as non-scientific. Perhaps a more acceptable term can be used to describe what is already a scientifically-accepted practice [Wil90] of rejecting outlying outcomes because the laws of statistics and probability confirm that those outcomes are infeasible, not representing reality. The probability cut-off varies (for example – any outcome not within 3 standard deviations from the mean of the other outcomes), but most fields [ERP03] of science have procedures for determining feasibility and infeasibility, even if that terminology is not used.

Some have suggested using Borel's guideline *"to set at 10^{-50} the value of negligible probabilities on the cosmic scale"* [Bor50]. Lloyd [Llo02] estimates that the Universe could contain no more than 10^{90} quanta (Note: many physicists believe that even a vacuum is filled with energy "quanta" that exceed the number of atoms by billions of times). Those quanta could be involved with no more than 10^{120} operations. Since the fastest chemical reaction known takes 10 femtoseconds (10^{-14} sec) [Zew99], if all 10^{80} atoms of the Universe participated in reactions of that speed for 14 billion years, less than 10^{111} reactions would theoretically take place. Since the vast majority of reactions are many orders of magnitude slower, if a scenario requires over 10^{111} reactions to become probable, that scenario is clearly infeasible within the Universe.

In a recent peer-reviewed paper, Abel notes, *"combinatorial imaginings and hypothetical scenarios can be endlessly argued simply on the grounds that they are theoretically possible. But there is a point beyond which arguing the plausibility of an absurdly low probability becomes operationally counterproductive"* [Abe09U]. He then calculates criteria and notes that the Universal Plausibility Principle (UPP) states that *"definitive operational falsification"* of any chance hypothesis is provided by inequalities based on the probabilistic resources of the Earth ($p < 10^{-70}$), solar system ($p < 10^{-85}$), galaxy ($p < 10^{-96}$), or Universe ($p < 10^{-108}$). If a scenario fails to meet the probability inequality standard, *"the hypothetical notion should be declared to be outside the bounds of scientific respectability. It should be flatly rejected as the equivalent of superstition."* For origin of life on Earth, any scenario with a probability less than 10^{-70} will be considered falsified scientifically (infeasible). Falsification is a criterion that would show a scientific theory is false if the criterion is shown to be true [Pop63] (see Appendix G).

Science needs a reality-check if origins are to be studied as science. This applies to the origin of mass and energy, the origin of life, and the origin of species. In what other science disciplines would outcomes be published as science if those outcomes had demonstrable probabilities of less than 10^{-100}? When "it's possible that ..." is used, scientists must verify that the pronouncement is indeed possible using known science, as opposed to really meaning "it may be speculated that..." Feasibility also needs to be verified using scientific principles described in the previous paragraph. The feasibility cut-off may vary depending on whether quantum, physical, or chemical interactions are involved, but there is a point where credulity is stretched beyond the breaking point, making science look "foolish" if persisting in treating such paths as pertinent. Without such safeguards, the public will be misled to believe something is science, as opposed to some scientist's speculation or belief. An educated public that understands the scientific meaning of "possible" and what "feasible" means should cause scientists to be more careful.

2 Information Basics: Data and Information Types

"The question 'How did life originate?' which interests us all, is inseparably linked to the question 'Where did the information come from?' Since the findings of James D. Watson and Francis H. C. Crick, it was increasingly realized by contemporary researchers that the information residing in the cells is of crucial importance for the existence of life. Anybody who wants to make meaningful statements about the origin of life would be forced to explain how the information originated" [Git97p99]. Carl Sagan wrote: *"The information content of a simple cell has been established as around 10^{12} bits, comparable to about a hundred million pages of the Encyclopaedia Britannica"* [Sag97]. That's in a "simple" cell!

This chapter will examine the basics of information in general and introduce how information applies to life. It's important to realize that information content is essentially massless, but the information medium has physical qualities. For example, a computer USB flash drive has the capacity to hold a certain quantity of data, but its weight doesn't measurably change by changing the data content.

The smallest numeric base that can hold data is binary since a bit (binary digit, also known as a Boolean datum) can be either 0 or 1. Anything that has only two possibilities can be represented by a bit, including resident/nonresident, married/single, male/female, etc., with an arbitrary assignment of which choice is 1 and which is 0. For example,if "010" represents married resident female, then "101" represents single nonresident male. A byte (8 bits) can hold eight binary data.

A base-4 digit has possible numeric values 0, 1, 2, and 3. The maximum digit in any base is one less than the base since 0 is always the smallest digit. This means 9 is the maximum decimal (base-10) digit, and 1 is the maximum binary digit. Since $4 = 2^2$, a base-4 digit can be expressed by the two-bit binary number 00, 01, 10, or 11.

The DNA genetic information system, with exactly four "letters" (specifying digits ACGT), is digital in base-4. Since $4 = 2^2$, the genetic information is relatively easy to represent in binary computers and to use in the equations of information theory (described in Chapter 5), by using two bits per nucleotide letter (details in Chapter 4).

"Information" has three significant meanings that are important when considering the information of life: "functional" [Szo03, Dur07], "Shannon" [Sha48], and "prescripti--ve" [Abe09P]. Information always involves contingency that rules out other possibilities. Data and Shannon information (probabilistic complexity) may or may not be useful. Functional information (a subset of Shannon information), on the other hand, is useful or meaningful (about something). Prescriptive information is an algorithmic subset (recipe) of functional information.

As an example to illustrate the three information types, consider the data typed into a word-processing program. Most such data is functional in that it has a purpose of communicating information to the ultimate reader of that information. If a monkey typed random data into the program, that complex data would have no purpose, but would have a very high Shannon information content since Shannon information deals only with the probability of the data pattern, irrespective of any meaning. A computer program typed into the wordprocessor is more than just functional, but is prescriptive in that it contains instructions to accomplish objectives based on data to be supplied during the execution of the program being typed. Prescriptive information expresses the decisions to be made and the criteria for the different execution paths. A computer problem is formally solved before physically implementing it (a program doesn't just appear on a disk).

To further clarify the difference between data and information, a "blank" disk drive has data (zeros and ones), but no functional information since the data are meaningless. If the disk were erased so that all data were zero, there would technically be one bit of Shannon information (plus a quantity if capacity is unknown) which could have been produced by the algorithm (prescriptive problem solution):
For each bit in the capacity(write 0). If the "0" in the algorithm were replaced by "random 0 or 1," there would be no functional information recorded since random (chance) data contains no functional information as it is unpredictable and functionally useless.

Any rational number (ratio of two integers) has information limited by the repeating portion of the division result. For example, π (pi) can be approximated by 22/7 = 3.$\underline{142857}$..., with the underlined digits repeating forever. If additional digits were written, those digits would be data, but would add no information. The exact representation of π cannot be expressed by any finite sequence of numbers since π is not rational. This does not mean that π has infinite information, since π can be calculated to any degree of precision desired by a computer program of finite length. The information in π is no more than the number of bits in such a program.

The difference between data and information can be illustrated with the difference between a protein of life and a simple chemical polymer. A protein has high functional information content since the sequence of its amino acid components is very specific, and would require an algorithm as long as the number of acids to express that information, according to algorithmic information theory [Cha07]. Polyethylene (multiple ethylenes), on the other hand, could be considerably longer than a protein (more data), but contains very little information as its algorithm (information can't exceed the number of bits to implement) could be:

write("H"); repeat write("-CH$_2$-") until randomly stopped; write("H").

Note that random data contains no functional information since there is no specificity. Exceptions would be games of chance, such as a lottery, which contains pseudo-information in that a winner is declared arbitrarily to be one picking a match to the randomly produced data.

Sometimes the information content may not be obvious, as in the following two 29-bit strings:

11011100101110111100010011010

and 10101100001110111011101000011.

The first string is counting to ten: 1 10 11 100 101 110 111 1000 1001 1010, whereas the second string was generated by a binary random number generator. If each bit were reversed, the first would still contain the same information, but in an encrypted form, making the information harder to ascertain. Note that the probability of each pattern is $2^{-29} = 1/536870912 = 1.86 \times 10^{-9}$. Note also that the first string isn't necessarily functional information since it may be feasible to generate it by random processes, and random processes can never produce functional information.

Functional information requires that the sender and intended receiver (which could be the same as the sender for information storage) of the data agree in advance as to the protocol and meaning of the information. For example, a single bit of information could be conveyed by the protocol according to the color of a blouse following a proposal: red means I'll marry you, blue means I won't. Other clothing colors would be "noise" to be ignored until the meaningful binary datum is communicated. The "one if by land, two if by sea" is also a binary message using two unary (base 1, which can't store information) objects.

If the intended receiver fails to receive the information, the message remains information. For example, if an encrypted message is intercepted, considerable time and effort may be spent trying to extract the information, even if never successful (the interceptor wasn't the intended receiver).

The hand can be an example of an information medium. There are over 1000 signs defined by American Sign Language [Ten98], but simpler finger communication is widely used by most people. The most widely used system is the five (or ten) unary object protocol, holding up the number of digits to represent a number. One could use each digit as a binary number so that 1024 (= 2^{10}) values could be represented using ten fingers, but most people would find that protocol awkward.

In order for meaningful communication to occur, both sender and receiver must know the protocol and meaning. Note that it is not quantity of message that makes information, it is the agreed protocol. That protocol involves arbitrary rules, not laws. Nothing forces a protocol.

The SETI (Search for Extraterrestrial Intelligence) project [SETI] is the largest single (distributed) computer application in the world, executing about 4×10^{13} floating-point operations per second (operations on non-integers). It uses over three million computers connected over the Internet to analyze data (from radio telescopes that listen to narrow-bandwidth radio signals from space) attempting to extract information that might indicate intelligent life is somewhere other than Earth. If one were to detect the previous 29-bit pattern (page 9) within the analyzed signals, one wouldn't publish the finding as proof that there are aliens out there since that pattern is probable approximately every two hours on each computer. If, however, one were to detect the first 100 prime numbers (number divisible only by itself and 1) in the data, the infeasibility (probability is 10^{-233}) of that pattern arising by chance would indicate that intelligence is out there.

If data from each of the maximum estimated 7×10^{21} stars [Ast03] were examined every 10^{-9} second over 14 billion years, the probability of even one such pattern occurring by chance is 10^{-183}, which is clearly infeasible. In the movie and book "Contact" [Sag85], intelligence was detected in 25 prime numbers found.

Note that unambiguously detecting information without an agreed protocol requires detecting meaningful data of such quantity that chance is eliminated as its cause. For example, if E.T. sent a one-time sequence of the first ten prime numbers, that sequence would be information from the sender, but couldn't be differentiated from random data that just happened to have the same value as a meaningful message had there been an agreed protocol for transmitting a short prime number sequence. For a small amount of data without an agreed protocol, information cannot be detected; at best, one could detect a pattern that may be information. Note that once a pattern is known "for certain," additional data aren't informational since they are predictable (a certainty isn't information).

A code that defines the meaning of a message symbol or combination of symbols (semiotics) is often used to specify functional information. In a zip code, for example, there is nothing intrinsic in a particular string of 5 digits that causes it to refer to a particular town or portion of a city. It has been established by the Postal Service to have the coded meaning ascribed to it. Those five digits can be decoded during mail processing so the item is delivered to the correct post office. Using zip+4 encoding, delivery can be made to the specific address during decoding.

The American Standard Code for Information Interchange (ASCII) defines the character represented by a particular 7-bit pattern, and is the used for text storage and communication. Imagine the chaos if each manufacturer chose its own arbitrary code to represent characters. What would display if "1100101" were sent to a computer? That particular pattern means 'e' using ASCII, with each other character having its own code. The ASCII standard allows meaningful communication messages to be transferred between equipment programmed to use that standard.

Life has considerable information, including multiple coding systems, in the interacting processing systems, the memories that store its programs and data, and the communications media. Chapter 4 provides a brief introduction to life, highlighting the informational aspects of life. Richard Dawkins has noted, *"The machine code of the genes is uncannily computer-like. Apart from differences in jargon, the pages of a molecular biology journal might be interchanged with those of a computer engineering journal"* [Daw95p17]. The different forms of life's information will be expounded in later chapters, demonstrating how that information is implemented in the components of life. The next paragraphs preview how the different forms of information may be applied to life.

Functional information principles may be used to quantify functional information by calculating a value which *"represents the probability that an arbitrary configuration of a system will achieve a specific function to a specified degree"* [Haz07]. Life is filled with information related to specific functional purposes. When the functional information for the required functions of life are considered as a whole, the figures become quite daunting. For example, Kalinsky uses functional information equations to show the simplest life form to be $10^{80,000}$ times more probable to have a controlled source than an accidental source [Kal08].

Prescriptive information (PI) *"instructs or directly produces nontrivial function... Prescriptive information either tells us what choices to make, or it is a recordation of wise choices already made"* [Abe09P]. PI involves formal (non-physical) choices at decision points that cannot be generated by chance or necessity (law). The information in life is fundamentally formal, not just physical. Life's information is stored in the physical DNA medium and other structures, but there is no known natural explanation as to how choices were explicitly made for the nucleotide sequence in order for DNA to contain the PI that is clearly evident, as covered in chapters 4 and 6.

Shannon information theory [Sha48] deals with the reduction of possibilities or uncertainty. The amount of information in a string of symbols is inversely related to the probability of the occurrence of that string. The more improbable the string, the more uncertainty it reduces and the more "information" it conveys. Probabilistic complexity (a better term than "information") provides only a mathematical measure of improbability, not whether a symbol string is meaningful or significant.

This lack of any meaning for Shannon uncertainty makes it inappropriate to be considered as the only "information" within life, since life not only has complexity, but also functionality. Some have extended Shannon "information" ("information" will be used when complexity is quantified) by incorporating a functionality variable to produce "Functional Sequence Complexity" in units of Fits (functional bits) [Dur07].

Information theory is concerned with storing and transmitting data in a manner that ensures the integrity of the data, regardless of the meaning attached to that data. For example, copying a file that contains 10^6 bits of random data (zero functional information) would be done with the same fidelity as copying a file of the same length containing financial records. For those two cases, it's very likely that the random data has a higher Shannon "information" content, since the probability of the random data pattern is probably lower than that of the financial data. The financial data probably has many repeated patterns, which would not be the case for the random data. Repeated patterns add little or no additional Shannon or functional information, but do add additional data, increasing the length of a data sequence.

3 Evolution of Computer Hardware and Software

Babbage's Son's Analytical Engine 1910 Model [Bab1910]

In order to understand the computer aspects of life, it is useful to examine how computer hardware and software have been developed. This may expand your concept of "computer" and memory beyond what has become so common. Most of the information in this short chapter results from over 20 years of teaching computer science.

The "Father of Computers" is Charles Babbage, who first described his Analytical Engine in 1837. Because of the complexity and cost of this mechanical computer, along with lack of good project control, Babbage was unable to successfully construct it. His son finished the scaled-down version pictured in 1910. The Engine did have all the components (program, processor, storage, input, and output) of modern computers, which were implemented in electronic computers over 100 years after Babbage's concept. The engine was "Turing-complete," which means it can theoretically be programmed to perform any computation that any machine can compute. Such a machine has at least memory and conditional branching for choosing different execution paths based on current data. It is important to realize that a computer needn't be electronic. In 1843, Ada Lovelace specified in complete detail a method for calculating Bernoulli numbers (Appendix A) with the Engine, which is recognized as the world's first computer program.

The first operational electro-mechanical computer was the Zuse in 1941 (Germany). It lacked a conditional branch needed to implement a "Turing-complete" computer (except [Roj98] by using an impractical program tape long enough to compute every possibility, ignoring all but the desired result). The Atanasoff–Berry Computer, built in 1942, was not Turing-complete and couldn't be reprogrammed. In 1973, a U.S.

District Court (Case 4-67 Civil 138, 180 USPO 670) concluded that the ABC was the first "electronic computer," thus putting the invention of the electronic digital computer into the public domain since its development was done using public funds.

The ENIAC (Electronic Numerical Integrator And Computer), operational in 1946, was the first general-purpose and Turing-complete electronic computer. It was programmable by patch cords and switches, using punched cards for input and output. Its memory and processing unit used vacuum tubes, consuming 150 kW of power. The ENIAC had ROM (read only memory) added in 1948. Core memory, which stores data in magnetic "donuts," was added to ENIAC in 1952. Core was the dominant main memory from the mid-50's through early 70's.

The Mark 1 computer (1949) used Williams cathode ray tube memory (which could serially store 500-1000 bits) and magnetic drum memory (similar to today's disks, without the need to move the read/write head). The UNIVAC I (UNIVersal Automatic Computer I) was the first US commercial computer in 1951. Its main memory used mercury delay lines, in which electric pulses were transduced into mechanical waves that propagated relatively slowly through a cylinder filled with mercury (this was pre-EPA).

Until 1955, when FORTRAN (FORmula TRANslator) was developed, programming was accomplished at a machine-code (native language) level or a one-to-one code translator known as assembly language, requiring programmers to think as the computer hardware functions. As programs became more and more complex, it became increasingly difficult to create them without significant "bugs" since the entire solution couldn't be visualized at once.

In the late 50's, operating systems (OSs) began to be introduced. An OS is a resource manager for a computer system, relieving the programmer from directly needing to do low-level operations (such as input and output). OSs became increasingly important as computer resources were shared, such as data files on a disk. The protocol for invoking OS functions allowed communication between the application program and the OS. When OSs were developed to execute on different types of computers using the same protocols, programs could be written without a concern for the actual computer's hardware. For example, the development of UNIX in the early 70's permitted programs to be executed on any UNIX system, regardless of its underlying hardware.

Starting in the late 60's, a shift was made toward structured programming and software engineering (SWE) with top-down design being a major emphasis. A problem was solved algorithmically (the logical steps needed for a solution), beginning at the highest level, and step-wise refining the modules of the solution, until a refinement can be implemented in a programming language. SWE allowed programs to be implemented, tested, and maintained utilizing hundreds of programmers writing millions of instructions. Bottom-up programming (starting at the lowest level, putting the parts together to get a working whole, was dismissed as too error-prone to be feasible for anything requiring more than a few hundred instructions). SWE requires that a working complex program is designed and implemented top-down since the modules must agree on the protocol of any incoming and outgoing data, as well as the specific functionality of each module.

Electronics technology advanced from transistors in the late 50's to integrated circuits (ICs) in the early 60's. In the mid 60's, minicomputers were introduced, and the large computers became known as "mainframe" computers. Concurrent with SWE, main memory moved from core to silicon RAM (randomly accessible memory, to differentiate this type of memory from disk memory, which is only block-accessible). Control store (microcode), that interprets the machine language code to produce control signals, became popular in the 70's.

In 1971, the Intel 4004 was implemented as the first microprocessor, with the CPU (central processing unit – the "brains" of a computer) on a single silicon IC chip (a business calculator used the 4004). The first general-purpose microprocessor-based personal computer (PC) was the Altair in 1975. It came as a "build-it-yourself" kit that used the Intel 8080 microprocessor. Apple introduced a pre-built PC in 1976, with other manufacturers soon following. The early PCs (including the IBM PC released in 1981) used cassette tape as the standard auxiliary storage. By the late 70's, floppy disks could be added to PC systems so that a PC could have an operating system. The vast majority of microcomputers today are built into appliances or controllers/monitors (e.g. in automobiles, microwave ovens, remote controllers, etc.). These special-purpose applications typically have programs and constants in ROM, have a limited set of input/output devices, and do not have a disk or OS.

At the same time that PCs became available to expand the low-end computers, the Cray-1 supercomputer was introduced (1976) to expand high-end computing (other less-successful supercomputers preceded the Cray-1). Other multi-computer systems, with appropriate parallel-processing modes were developed extensively starting in the early 80's. These complex systems, in which multiple computers could simultaneously access shared data in memory, introduced new levels of complexity in programs and operating systems. SWE principles had to expand to accommodate this increased complexity as "bugs" could occur that had no direct connection to the program that was executing in a particular processor (e.g. by failing to synchronize correctly). Having access to distributed memory, especially NUMA (non-uniform memory access), required ever more complexity in protocols for reliable data communication, which has been a primary research interest of this author [Joh94, Joh95, Joh97B, Joh97T, Joh05].

Over time, the trend has been a doubling of computer power, RAM capacity, disk capacity, and performance cost ("bang per buck") every two years. Hardware and software engineers will undoubtedly continue the trend for the foreseeable future. An interesting observation is that one metric of the cost of computers has remained fairly constant for the past 60 years – about $100 per pound.

Biologically-inspired computer systems began in 1994 [Adl94], with potential Turing-completeness later proven [Bon96]. Enzyme/DNA computers were proposed in 2002 [Lov03] and implemented in 2004 [Ben04]. A distributed computing system has been designed, patterned after life's computing systems [Bab06], and a massively parallel computing system was designed based on biological cells [Ban10]. Eventually, computer engineers may be able to harness the tremendous processing power exhibited in life. Computer hardware [D'On10] (see Appendix B) and operating systems [Yan10] comparisons have also been made with life's processing systems to find functional equivalences between the engineered and biological systems.

This short chapter highlights several aspects of computers and computer history that may be unfamiliar to many today. Computers may or may not be Turing-complete. They may be mechanical, electro-mechanical, electronic, or biological. Main memory may be sequentially or randomly accessible, and may or may not be shared. Languages can be translated or interpreted (including machine-language interpretation using microcode). Each incremental improvement in hardware or software involved engineering, including new SWE principles to exploit new hardware capabilities. Non-trivial programs require software engineering principles, including top-down design and implementation.

4 Life Basics

Although there is no universally accepted definition of life [Emm97], it often includes characteristics like metabolism, growth, adaptation, and reproduction. *"The existence of a genome and the genetic code divides living organisms from nonliving matter"* is perhaps the most concise definition of life [Yoc05p3]. This definition includes as living (or at least as once living) those that are sterile (e.g. mules and worker ants) and those not having cells (e.g. viruses, which aren't autonomous life). While life uses the laws of chemistry and physics, those laws cannot define or explain life any more than the rules of grammar that were used during the preparation of this book define its content.

This chapter presents the simplified basics of all life, and also includes a brief view of multicellular organisms. The very important terms (used in future chapters) that will be described include life's amino acid building blocks to make proteins (including enzymes), DNA/RNA with nucleotide/base sequences, codon, ATP, ribosome, genome, and gene. Other terms will be used, but needn't be understood in order to obtain a comprehension of and an appreciation for the cybernetic complexity of life. This short overview is not meant to make the reader a biologist, so don't get "stuck" on the technical details, but do appreciate how detailed life is and appreciate its functional complexity. You can return as needed to gain additional understanding.

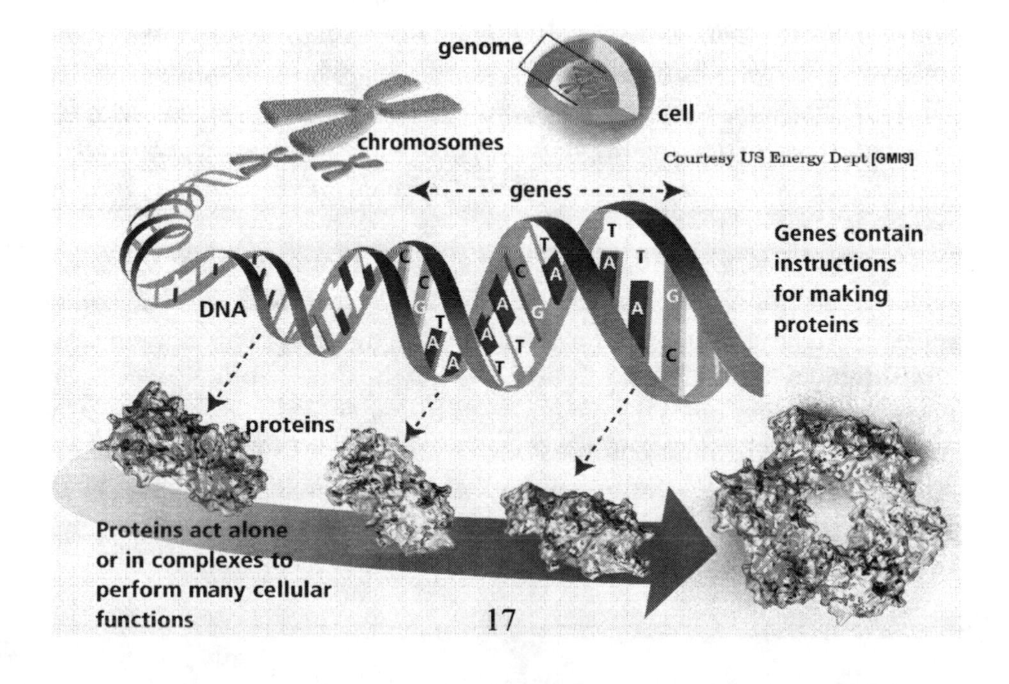

DNA (Deoxyribonucleic acid) contains the **genome** (genetic information) of a cell, including the information for constructing **proteins**, RNA (Ribonucleic acid), and other cellular components. The sequence of the four **bases** (ACGT) used in DNA's **nucleotide** sequence specifies an organism's genome. The base pairing between the two strands of the DNA helix is always G-C or A-T. DNA's information is often organized as **genes**, with a gene often specifying the protein to be manufactured. **Ribonucleic acid (RNA)** is similar to DNA with ribose replacing deoxyribose and U replacing T in a nucleotide's position along the nucleic acid sequence. See appendix C for more on life's structures.

A **protein** is a sequence of **amino acids** in a functional chain. **Enzymes** are catalytic proteins that have special slots to hold other molecules to make chemical reactions feasible. The slots include two for holding the reacting chemicals (e. g. two amino acids) , one for ATP (the main source of stored energy in cells), and slots for establishing which non-chemically-determined DNA/RNA codon (described shortly) is specified. Each enzyme enables a specific required chemical reaction without ultimately being changed itself. Life both requires and manufactures these enzymes, as well as all the other proteins.

Life's enzyme catalyzed reactions take place in the millisecond time-frame, whereas *"uncatalyzed reactions span a range of at least 19 orders of magnitude"* [Lad03]. The slowest uncatalyzed reaction takes over 10^{19} times as long as the fastest reaction. The longest known biochemical *"half-time - the time it takes for half the substance to be consumed - is 1 trillion years, 100 times longer than the lifetime of the universe. Enzymes can make this reaction happen in 10 milliseconds... Without catalysts, there would be no life at all... It makes you wonder how natural selection operated in such a way as to produce a protein that got off the ground as a primitive catalyst for such an extraordinarily slow reaction"* [Wol03].

The transcription process creates a coded message from the DNA code. The translation process builds a protein based on that coded message. These processes for manufacturing a protein are shown (at a simplified level) in the next figure. During the transcription process, an enzyme computer reads the genetic code by copying stretches of partially unwound DNA from a chromosome (a structure of DNA and proteins which keeps the DNA from tangling) within the nucleus into the related messenger RNA (**mRNA**). Over 30 proteins are used during transcription, including RNA polymerase (which is itself built from over 3,000 amino acids by the same transcription/translation processes). The mRNA has a base-paired (C-G, G-C, T-A, A-U) copy of the original DNA code, and is edited by other proteins to reduce errors. Note that mRNA is a communication medium, equivalent to a wire carrying electronic signals, which moves out of the cell's nucleus through the nuclear pore complex (NPC) into the cytoplasm, and ultimately to a ribosome.

Transfer RNA (tRNA), floating in the cytoplasm, has been programmed to pick up a particular amino acid, so that the **codon** specified by the particular type of tRNA will be associated with the required amino acid. The genetic code consists of three-letter codons formed from a sequence of three nucleotide bases (e.g. AAG or GAC). A **ribosome**, a complex of RNA and proteins, is a computerized protein-building factory that reads the mRNA sequence (using over 50 proteins and additional RNA). The ribosome has been programmed to fetch (via base-pairing) the particular tRNA that is specified by the current mRNA codon so that the specified amino acid molecule is carried to the protein assembly point during the translation process within the ribosome. The amino acid attached to the tRNA is thus transferred to the protein being constructed so that a specified sequence of amino acids results.

The linear algorithmic sequence of a gene's bases in DNA determines the sequence of amino acids in the protein, which in turn determines the properties of that protein. Those properties are largely

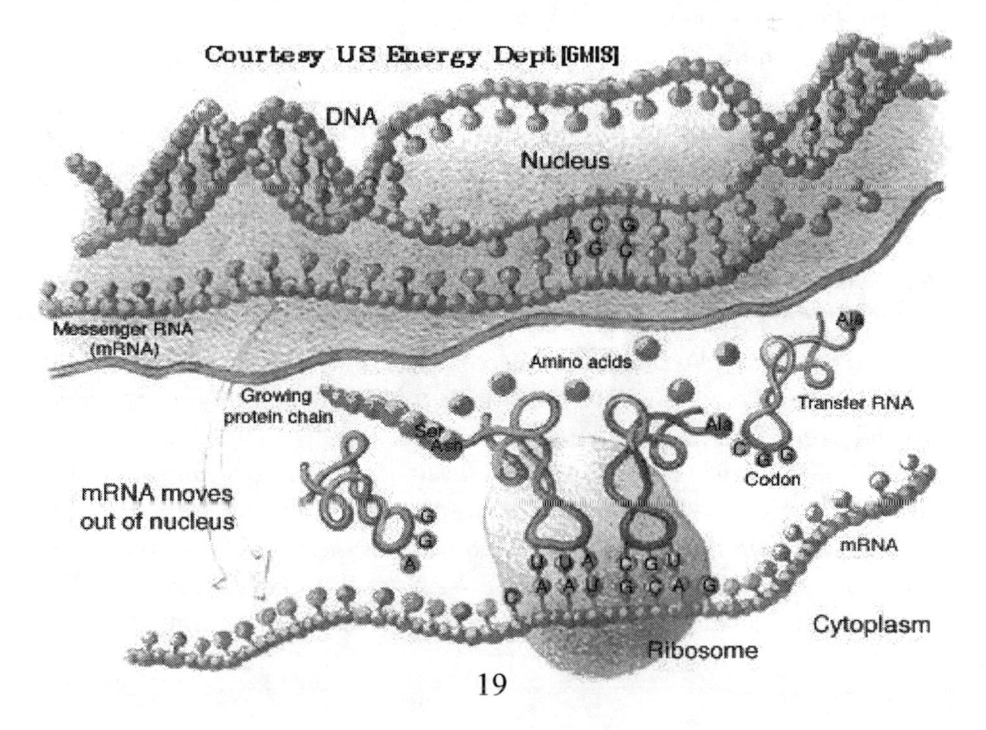

determined by the protein's three-dimensional structure, as the protein folds (assisted by other proteins) into a very complex functional physical structure. Changing a single amino acid can (but doesn't always, since some positions can tolerate an alternate amino acid) make the protein totally useless since how it folds depends on the specific sequence of amino acids.

Note that each of the over 150 proteins used during the protein-construction process is itself manufactured using the same type of process. All of these proteins, as well as the required DNA code, are required for manufacturing a protein. Proteins in the NPC (nuclear pore complex) are not only communication channels that regulate the passage of all molecules (e.g. mRNA) to and from a cell's nucleus, but also *"play a role in the organization of the genome and a very direct role in gene expression"* [Sal10] (genes can be turned on or off). The NPC probably also does proof-reading on mRNA while passing through [Gru10]. ATP is required to supply the energy to the enzymes so the required chemical reactions can take place, but its production not only requires 10 specific proteins and specific DNA code, but also requires ATP! Since life requires fully-functional ribosomes, DNA, RNAs, ATP, enzymes, other activator proteins, and a multitude of other cellular components, there have been no feasible purely physical scenarios proposed for the origin of this complex interactive cybernetic system. *"Nobody knows how it happened but, somehow, without violating the laws of physics and chemistry, a molecule arose that just happened to have the property of self-copying – a replicator"* [Daw96Cp282-3].

The question *"How did life begin?"* is one of the *"biggest unanswered questions"* in biology [New04]. *"More than 30 years of experimentation on the origin of life in the fields of chemical and molecular evolution have led to a better perception of the immensity of the problem of the origin of life on Earth rather than to its solution"* [Dos88]. *"What creates life out of the inanimate compounds that make up living things? No one knows. How were the first organisms assembled? Nature hasn't given us the slightest hint. If anything, the mystery has deepened over time"* [Eas07].

"The received view, today, is that life is but an extremely complex form of chemistry... The problem of which molecules came first has been the object of countless debates.... What really matters is that spontaneous genes and spontaneous proteins had the potential to evolve into the first cells. This however, is precisely what molecular biology does not

support. The genes and proteins of the first cells had to have biological specificity, and specific molecules cannot be formed spontaneously. They can only be manufactured by molecular machines, and their production requires entities like sequences and codes that simply do not exist in spontaneous processes. That is what really divides matter from life. All components of matter arise by spontaneous processes that do not require sequences and codes, whereas all components of life arise by manufacturing processes that do require these entities. It is sequences and codes that make the difference between life and matter. It is semiosis [symbol translation system] *that does not exist in the inanimate world, and that is why biology is not a complex form of chemistry"* [Bar08S].

The "RNA world" [Woe67, Cri68, Bad04] has been proposed to circumvent the intractable difficulty of accounting for the origination of life based on DNA/RNA/proteins, even though it's the only form of life actually known. In this scenario (currently the most widely accepted scenario), RNA functions as both an enzyme and as a replicator [Kru82, Joy98]. The discovery of ribozymes, the Nobel prize [Nob89] winning discovery that RNA can sometimes function as an enzyme, has spurred much research and speculation for the RNA world. *"The problem of the origin of life is the problem of the origin of the RNA World, and that everything that followed is in the domain of natural selection"* [Org04].

One of the biggest problems for the pre-RNA World model is finding sequences that can simultaneously self-replicate and catalyze the large number of needed metabolic functions, with all needed components arising at the same place at the same time. *"Little empirical evidence exists to contradict the contention that untemplated sequencing is dynamically inert (physically arbitrary). We are accustomed to thinking in terms of base-pairing complementarity determining sequencing. It is only in researching the pre-RNA world that the problem of single-stranded metabolically functional sequencing of ribonucleotides (or their analogs) becomes acute. And of course highly-ordered templated sequencing of RNA strands on natural surfaces such as clay offers no explanation for biofunctional sequencing. The question is never answered, 'From what source did the template derive its functional information?' In fact, no empirical evidence has been presented of a naturally occurring inorganic template that contains anything more than combinatorial uncertainty"* with no functionality [Abe05]. A template could theoretically hold information, but there's no feasible source known for that information (though several infeasible scenarios are proposed).

Other speculation (e. g. osmosis-first and metabolism-first) of life's origin attempts to look outside the mainstream views. *"Despite thermodynamic, bioenergetic* [metabolism] *and phylogenetic* [evolutionary relatedness] *failings, the 81-year-old concept of primordial soup remains central to mainstream thinking on the origin of life... Here we consider how the earliest cells might have harnessed a geochemically created proton-motive* [electrochemical energy] *force and then learned to make their own, a transition that was necessary for their escape from the vents"* [Lan10]. This is basically an "osmosis-first" (chemical migration through membrane) theory of origins.

"The replicator concept is at the core of genetics-first theories of the origin of life, which suggest that self-replicating oligonucleotides [DNA/RNA] *or their similar ancestors may have been the first "living" systems and may have led to the evolution of an RNA world. But problems with the nonenzymatic synthesis of biopolymers and the origin of template* [as an information source] *replication have spurred the alternative metabolism-first scenario, where self-reproducing and evolving proto-metabolic networks* [the first chemical cycles] *are assumed to have predated self-replicating genes... In sharp contrast with template-dependent replication dynamics, we demonstrate here that replication of compositional* [composome = "bag" of self-replicating chemicals] *information is so inaccurate that fitter compositional genomes cannot be maintained by selection and, therefore, the system lacks evolvability (i.e., it cannot substantially depart from the asymptotic steady-state solution already built-in in the dynamical equations). We conclude that this fundamental limitation of ensemble replicators cautions against metabolism-first theories of the origin of life"* [Vas10]. Without the information in a genome (or equivalent), evolution can't happen, so that composomes are ruled out as the first "life."

Since there is no known scientific procedure to generate life in the laboratory, let alone by some unknown prebiotic mechanism, one could assume the probability of life from purely physical causes is zero. What often is assumed is that since life obviously exists and the only allowable mechanism is physicality, it must have occurred that way, despite the improbability. That is an obvious tautology since the proposition "life can only be the result of purely physical sources" uses that proposition to prove itself (the proposition that "life exists" is obviously true). Two-time Nobel Prize winner Ilya Prigogine noted something that has never been proved incorrect, *"The statistical probability that organic structures*

"If I could just DO this, I'll have proven that life arose by undirected natural processes!"

and the most precisely harmonized reactions that typify living organisms would be generated by accident, is zero" [Pri72]. See Appendix C for additional OOL information.

As far as science knows, the law of biogenesis, life only arises from life, is still valid. Any statement that begins, "It's possible that life originated from non-life by ..." is misstated from a probability point of view since non-zero probability has never been proved. In fact, the low probabilities of the scenarios proposed thus far have operationally falsified them [Abe09U] (Appendix G has several required falsifications). In addition, it seems impossible (zero probability based on information sciences) for life's information to have originated by inanimate physical science natural interactions, as will be discussed in chapters 6-7. Science is in need of real breakthroughs to resolve these issues.

Helicases (enzyme proteins) are molecular motors that use the chemical energy of ATP to break hydrogen bonds between bases and unwind the DNA double helix into single strands [Tut04]. *"DNA helicases act as critical components in many molecular machineries orchestrating DNA repair in the cell... Multiple diseases including cancer and aging are associated with malfunctions in these enzymes... Helicases are a special category of molecular motors that modify DNA ... by moving along strands of DNA, much the same way cars move on roads,* *using an energy-packed molecule, adenosine triphosphate (ATP) as a fuel source. Their primary function is to unzip double-stranded DNA, allowing replication and repair of the strands"* [Spi08].

Biology professor Jerry Bergman [Ber97] uses a good analogy for the DNA replication process during cell division to create two cells from a single cell. Since all life starts as a single cell, this description applies to every organism. Scale-up DNA by a factor of one million to the equivalent of two 125-mile long strands of fisherman's monofilament line wrapped together to form a double-helix structure, neatly folded and packed to fit into a basketball (nucleus equivalent). Envision the engineering problem of creating an exact duplicate of each 125-mile long line to form two identical sets of those pairs of twisted lines, each packed into a new basketball! During cell division, the entire length of DNA must be split apart, duplicated, and repackaged for each daughter cell. There are about 25 million protein spools around which the DNA is wrapped, organized into an extremely complex hierarchical set of protein structures. Once an initiator protein locates the correct place to begin copying, a helicase "unzipper" unwinds the strands at approximately 8000 rpm, forming a fork area, without tangling the DNA strands as they separate. An "untwister" enzyme (topo-isomerase) systematically cuts and repairs resulting strands to prevent tangling as each DNA strand is formed. Other enzymes copy the flat, untwisted sections of DNA, which

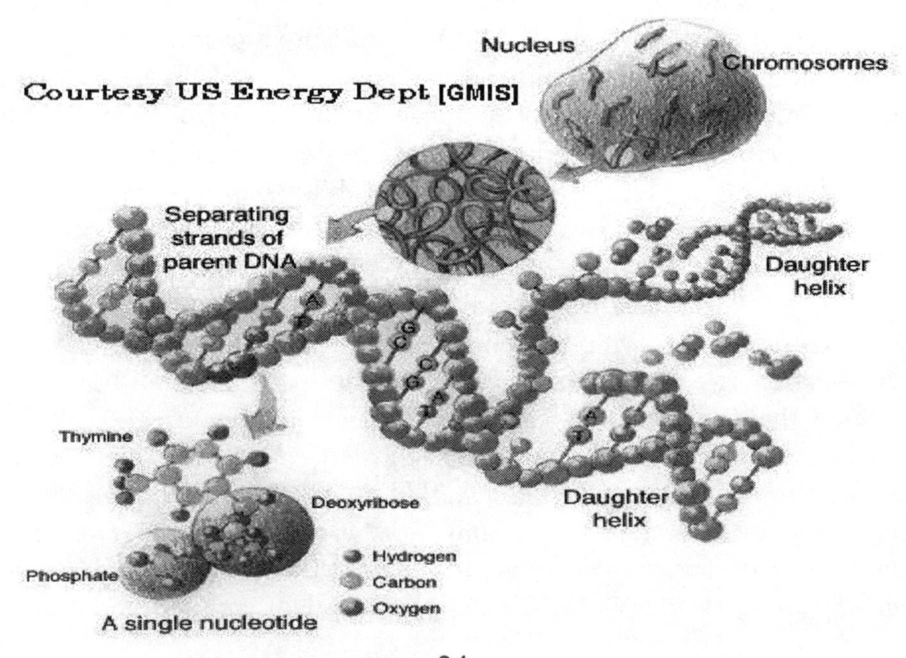

are then connected together via DNA ligases into one continuous strand. There are over 30 specific functional proteins required for cell replication, each manufactured according to its own implemented computer algorithm (chapter 6 covers this in detail), with all processes digitally controlled.

It has also been recently discovered that the replication protocol has higher priority than the protein-manufacturing protocol [Roc10]. The replisome runs along the same path as the RNA polymerases (for protein transcription), but causes any polymerase to abort its task so that replication can be done reliably.

The DNA sequence is digital in base-4 digits since there are four bases (note that "base" can be numeric or a chemical in a nucleotide). The digits A, C, G, and T are equivalent to the first four decimal digits 0, 1, 2, and 3, but at this point it is unknown which base identifier has which value if there is a difference (currently, the assignment is arbitrary). Since each codon is a 3-letter combination, there are 4^3 (= 64) possible codons. This is equivalent to the 10^3 (= 1000) different 3-digit (0-9) decimal numbers that can be formed. The codons encode the twenty standard amino acids (1-letter in the following table), giving most amino acids more than one possible codon, with codes left over for 3 "stop" codons ("$" in table) signifying the end of the gene-coding region space.

Non-standard amino acids are sometimes substituted for standard stop codons, depending on the specific mRNA sequence. There is considerable speculation and important findings on the "non-coding" parts of DNA, often dismissed as "junk DNA." These parts include intragenic regions (DNA code between genes) known as **introns**. The following 64 codons produce the amino acids shown during protein building (ATG* also starts transcription if not already in a gene).

TTT P	TCT S	TAT Y	TGT C	ATT I	ACT T	AAT N	AGT S
TTC P	TCC S	TAC Y	TGC C	ATC I	ACC T	AAC N	AGC S
TTA L	TCA S	TAA $	TGA $	ATA I	ACA T	AAA K	AGA R
TTG L	TCG S	TAG $	TGG W	ATG M*	ACG T	AAG K	AGG R
CTT L	CCT P	CAT H	CGT R	GTT V	GCT A	GAT D	GGT G
CTC L	CCC P	CAC H	CGC R	GTC V	GCC A	GAC D	GGC G
CTA L	CCA P	CAA Q	CGA R	GTA V	GCA A	GAA E	GGA G
CTG L	CCG P	CAG Q	CGG R	GTG V	GCG A	GAG E	GGG G

The smallest genome (though not autonomous) found so far is in *"the psyllid symbiont Carsonella ruddii, which consists of a circular chromosome of 159,662 base pairs... The genome has a high coding density (97%) with many overlapping genes and reduced gene length"* [Nak06]. Overlapping genes occur when the same nucleotides are part of different codons. For example, the genetic sequence "AGACATG" could have codons AGA & CAT or GAC & ATG, depending on whether a codon starts on the first or second nucleotide in the sequence. To illustrate the concept of overlapping genes, consider three short artificial "proteins," each having its own pre-coding enabling sequence (not specified here), generated by the following artificial genome sequence.

```
..ATGTGTGATGCTACCCTATGTCCAAAAGGGCACCTGCCAATAACCTAGGGGTGA
P1: M   C   D   A   T   L   C   P   K   G   H   L   P   I   T
p2:             M   L   P   Y   V   Q   K   G   T   C   Q
p3:                         M   S   K   R   A   P   A   N   N   L   G
```

Overlapping genes add complexity since *"Normally, transcriptional overlap can interfere with expression of a gene, but these genomes cope with high frequencies of overlap and with termination signals within expressed genes"* [Wil05]. Since the DNA code is like computer code in many respects, it is truly amazing that the same "instruction patterns" can perform different overlapping instructions. The author has done this with short sequences (up to eight bytes) of computer code in assembly language, and can assure the reader that it is not trivial to make meaningful operational sequences in which completely different instructions result by starting execution at different locations! If you doubt that isn't trivial, try to come up with a string of letters that would have different meaningful messages by starting at different letters of the string. To make it even harder, make each of the overlapping messages several hundred letters long as would be typical in a gene. DNA's information in is multi-leveled, with other more complex mechanisms for mRNA construction.

The human genome has thousands of overlapping genes. *"However, the origin and evolution of overlapping genes are still unknown"* [Vee04]. Human DNA contains an estimated 20,000–25,000 genes [Ste04] in its approximately 3 billion base pairs. This is down from earlier estimates of over 100,000 genes. *"The Human Genome Sequence Reveals Unexpected Complexity... Only 1.5% of the 3.2 billion base pairs of the human genome encode protein, yet those 31,000 or so genes specify 100,000 to 200,000 distinct proteins"* [Lew06]. *"At least part of the information for the extra proteins may come from the presence of hitherto undiscovered overlapping genes, although more may come from alternative splicing of exons* [tRNA or mRNA sequence] *in a single gene"* [Ans-Web]. *"Indeed, as a result of the overlapping genetic messages and different modes of information processing, the specified information stored in DNA is now recognized to be orders of magnitude greater than was initially thought"* [Mey09p462]. The DNA information storage medium is already considerably more complex than any other system known (including anything man-made), and we still don't know the extent of that complexity.

Recently, sub-coded information [Can10] and a second genetic code [Bar10] characterizing alternative splicing have been discovered. Various transcribed RNAs are mixed and matched and spliced into mRNAs for specifying protein construction and other controls, sometimes joining messages that were separated by thousands of nucleotides. *"For example, three neurexin genes can generate over 3,000 genetic messages that help control the wiring of the brain"* [Fre10]. MicroRNAs (small RNA segments) regulate large networks of genes by acting as master control switches [Lie10]. Tiny polypeptides (with 11-32 amino acids) can function as "micro-protein" gene expression regulators [Kon10].

Epigenetics [epi- means "above"] is a research field that has found that cell fate is governed by chemical changes to DNA and its associated proteins, not just by the genome. *"When epigenetic processes go awry, diseases may occur. Epigenetics is therefore emerging as an important research area with relevance to understanding many adult conditions like heart disease, diabetes, obesity, cancer and autoimmune disorders"* [Bab10]. In addition to control signals originating in genes, other controls originate in the non-coding portion of DNA, and also in the epigenome.

Control is an extremely important aspect of life that will be covered in more detail in chapters 6-7. For example, the chemistry of life is very controlled, as illustrated by "handedness." The left and right hands have the same structural elements, but are mirror images. Amino acids and ribose/deoxyribose sugars also exhibit handedness (see appendix C), with only one form (left-handed amino acids, right-handed sugars) being useful or generated in life. A single instance of a molecule of the incorrect handedness renders the complete structure useless. This differs greatly from organic chemistry, where each form is 50% of the product. Life uses chemistry, but that chemistry is controlled by life's cybernetics.

Recall that both the sender and receiver of information must know the message protocol for information to be communicated successfully. This includes the discovery that *"proteins communicate not by a series of simple one-to-one communications, but by a complex network of chemical messages"* [Edi10]. So far, no credible origin of any code by inanimate nature has been proposed. Explaining the origin of codes within codes will be even more difficult, as this is a multi-level encryption system. This is equivalent to a spy communicating using a hidden message within an encrypted message – e.g. using the third letter of each word as the first level, use the second letter of the word preceding the first-level word in a known dictionary as the final message letter.

The human genome has 23 pairs of chromosomes, which are organized structures of DNA and proteins found in cells. If stretched out, DNA would be a six-foot chain weighing about a trillionth of a gram. One DNA molecule from each human that has ever lived [Peo94] would have a total mass under 0.1 gram. If all the DNA in a human's body were laid end-to-end, it would be about 30 billion miles long [Spe97p30], over 320 times the distance from the Earth to the Sun.

DNA is extremely information rich, as Richard Dawkins writes, *"Biology is the study of complicated things that give the appearance of having been designed for a purpose... Physics books may be complicated, but ...The objects and phenomena that a physics book describes are simpler than a single cell in the body of its author. And the author consists of trillions of those cells, many of them different from each other, organized with intricate architecture and precision-engineering into a working machine capable of writing a book"* [Daw96Bp1-3].

One might think that the genetic code is "inefficient" in that a codon has 64 possibilities, with only 20 amino acids being specified. It may seem that a different alphabet than ACGT, perhaps expanded, would be preferable. It has been discovered that the DNA alphabet is optimum: *"the best of all possible codes"* [Fre00]. Recently, researchers have *"boosted the number of amino acids that can be built into a protein from the 20 covered by the existing genetic code to 276. That's because Chin's new code creates 256 possible four-letter nucleotide words or 'codons,' each of which can be assigned to an amino acid that doesn't currently exist in living cells... Chin's team redesigned several pieces of the cell's protein-building machinery, including ribosomes and transfer RNAs (tRNAs). Together, they read the genetic code and match it up to amino acids"* [Ged10]. Some have questioned the safety of such genetic engineering since such unnatural organisms could have unpredictable severe consequences when interacting with Earth's form of life.

When a recently discovered chemical structural parity-bit *"error-coding approach is coupled with chemical constraints, the natural alphabet of A, C, G, and T emerges as the optimal solution for nucleotides"* [Mac06]. This parity error-check can be examined by helicases during unzipping so that structures with errors may be destroyed rather than producing a defective structure. Parity will detect any odd number of errors (with a single error being the most common). In most cases, a single nucleotide error will produce the same amino acid when transcribing the gene. Only Methionine (the genetic start code) and Tryptophan have no redundant codes. *"The fact that more than one codon is assigned to eighteen of the more common amino acids in protein is seen as very natural, and indeed necessary, to achieve a moderate error-correcting capability in the genetic code"* [Yoc05p42]. Error correction [May04] and repair mechanisms make mutations very rare.

A typical cell (except prokaryotic which lacks a membrane-bound nucleus found in an eukaryotic cell) has a nucleus that contains the DNA, whereas the cytoplasm outside the nucleus contains specialized organelles ("organs" of a cell). Ernest Borek states, *"The membrane recognizes with its uncanny molecular memory the hundreds of compounds swimming around it and permits or denies passage according to the cell's requirements"* [Bor73].

Inside the lipoprotein cell membrane are specialized components to perform nutrition, repair, waste disposal, communication, and reproduction. Some organelles are extremely complex, consisting of dozens of proteins coordinated into working machines. One such organelle, the bacterial flagellum (a rotary motor/propeller assembly), will be discussed in some detail in chapter 8. The organelles include mitochondria that provide energy, the protein-producing factories called ribosomes, golgi that package and store the manufactured proteins, and lysosomes that dispose of waste. *"If genomic DNA is the cell's planning authority, then the ribosome is its factory, churning out the proteins of life. It's a huge complex of protein and RNA with a practical and life-affirming purpose-catalyzing protein synthesis. Bacterial cells typically contain tens of thousands of ribosomes, and eukaryotic cells can contain hundreds of thousands or even a few million of them"* [Bor07]. All that is in a single cell!

In multicellular organisms, cells are typically differentiated to perform specialized functions. Almost all organisms that can be seen without magnification are multicellular, as are all members of the Plantae and Animalia kingdoms. All organisms begin as a single cell, but multicellular organisms have been programmed to produce cells that have considerably different characteristics, even though the DNA is identical. *"Cell fate is governed not only by the genome, but also by chemical* *changes to DNA and its associated proteins, a research field called* ***epigenetics****. These 'epigenetic'* [epi- means above] *tags are one of the ways that genes get switched on or off in different places at different times, enabling different tissues and organs to arise from a single fertilised egg"* [Kim10]. The **epigenome** causes genes to be expressed differently. For example, a stem cell may change into one of many cell types (muscle, brain, blood vessels etc.) as it continues to divide, activating some genes while suppressing others. In early embryonic development, a series of chemical signals from different signaling pathways cause a single undifferentiated cell to become a highly specialized complex organism with a variety of different cell types arranged in very precise patterns. These patterns, ensure that all the body structures develop correctly, each in the appropriate place. Development uses a precisely regulated interplay of different cell types. *"It is fascinating how the genetic programme of an organism is able to produce such different cell types out of identical precursor cells"* [Sch10].

An example of a specialized organ is the brain, which can be extremely complex. The human brain contains about 100 billion neuron cells, each linked to as many as 10,000 other neurons with over 10^{15} total synaptic interconnections. It is estimated that the human brain can perform in excess of 10^{16} operations per second [Hor08], which is more than all the computers in the world put together (although that may not be true much longer due to the advances in computer power and widespread computer use). This means that the most powerful "electrical" computer system in the world resides in the top of your head!

For example, one optic nerve's data to the brain in one second would require hours to process on the world's fastest computer, but your brain processes that data in real time. *"Many neuroscientists do assume that, just as computers operate according to a machine code, the brain's performance must depend on a 'neural code,' a set of rules or algorithms that transforms those* [synaptic electrical] *spikes into perceptions, memories, meanings, sensations, and intentions*" [Hor08]. The brain uses 20% of the body's total oxygen consumption and 15% of the blood flow, even though it is only 2% of the body's weight [Rai02, Cal06].

"For me, the brain is not a supercomputer in which the neurons are transistors; rather it is as if each individual neuron is itself a computer, and the brain a vast community of microscopic computers. But even this model is probably too simplistic since the neuron processes data flexibly and on disparate levels, and is therefore far superior to any

digital system. If I am right, the human brain may be a trillion times more capable than we imagine, and 'artificial intelligence' a grandiose misnomer... I think it is time to acknowledge fully that living cells make us what we are, and to abandon reductionist thinking in favour of the study of whole cells. Reductionism has us peering ever closer at the fibres in the paper of a musical score, and analysing the printer's ink. I want us to experience the symphony" [For09].

The organs of a body work together to enable life for the body, including respiration, nutrition, waste disposal, repair, etc. *"In each of the some 300 trillion cells in every human body, the words of life churn almost flawlessly through our flesh and nervous system at a speed that utterly dwarfs the data rates of all the world's supercomputers. For example, just to assemble some 500 amino-acid units into each of the trillions of complex hemoglobin molecules that transfer oxygen from the lungs to bodily tissues takes a total of some 250 peta operations per second"* [Gil06]. (Hemoglobin production requires 2.5×10^{17} operations/sec!)

The **genome** is a unique set of programs embedded in the DNA memory for an organism. *"The great evolutionary biologist George C Williams has pointed out that animals with complicated life cycles need to code for the development of all stages in the life cycle, but they only have one genome with which to do so. A butterfly's genome has to hold the complete information needed for building a caterpillar as well as a butterfly. A sheep liver fluke has six distinct stages in its life cycle, each specialized for a different way of life"* [Daw98p24]. For example, the monarch butterfly has four life cycle stages: the egg, the larvae (caterpillar), the pupa (chrysalis), and the adult butterfly. It also has four generations of four different butterflies each going through all four stages during one year each, until it is time to start over again with stage one and generation one in the fifth year. The first generation starts in Mexico. The first three generations each live for 2-6 weeks as each generation migrates north. The fourth generation starts in Canada, and lives for 6-8 months as the butterfly flies 2,500 miles back to Mexico before laying the eggs for restarting the first generation. All of the information for all four generations, including the four stages of each generation, is part of the genome of the butterfly.

"The main distinctive features of the living beings are their extreme complexity, which is unmatched in the non-living world, and (not independently) the rather obvious but still overlooked fact that, besides

matter and energy, they receive and transmit information, and that life heavily relies on information transfer and conservation. This last point has no equivalent outside the living world and appears as the specific mark which radically differentiates the living world from the non-living one. It makes biology especially relevant to information theory... prompting biologists to use information theory as a main tool" [Bat07]. The next three chapters will investigate many more details of life's information.

Obviously, this chapter just scratches the surface of life. Many books have been written that go into much more detail on all aspects of this extremely complex topic. As more is known about life, the more intricate and complex it becomes. There is no such thing as a "simple" organism, as even the simplest organism is extremely complex, with information content and processing systems that dwarf anything humans have developed. *"The complexity of biology has seemed to grow by orders of magnitude... the more we know, the more we realize there is to know... it's infinitely more complex."* [Hay10] Life's description in this chapter will serve as a sufficient backdrop for the rest of this book.

5 Shannon Information in Life

Evolutionary biologist George Williams observed, *"Evolutionary biologists have failed to realize that they work with two more or less incommensurable* [incomparable] *domains: that of information and that of matter... These two domains will never be brought together in any kind of the sense usually implied by the term 'reductionism.'... Information doesn't have mass or charge or length in millimeters. Likewise, matter doesn't have bytes... This dearth of shared descriptors makes matter and information two separate domains of existence, which have to be discussed separately, in their own terms"* [Wil95].

This chapter will demonstrate how principles of information theory can be applied to life. Remember (chapter 2) that information can be stored only by contingency, that is, the storage digit can be placed arbitrarily into any of its possible values. If there were no contingency involved, the value would be determined by "law," and that digit would hold no information since a certainty cannot hold information. The four bases (ACGT) of DNA represent the four digits of the genetic numbering system (currently value assignment, such as A=0, is arbitrary).

"There are no chemical bonds between the bases. Thus, there are no chemical rules to determine the order in which the bases will be attached to the background" [Dav02]. *"One cell division lasts from 20 to 80 minutes, and during this time the entire molecular library, equivalent to one thousand books, is copied correctly"* [Git97p90].

"Crick expounded and enshrined what he called the 'Central Dogma' of molecular biology. The Central Dogma shows that influence can flow from the arrangement of the nucleotides on the DNA molecule to the arrangement of amino acids in proteins, but not from proteins to DNA. Like a sheet of paper or a series of magnetic points on a computer's hard disk or the electrical domains in a random-access memory – or indeed all the undulations of the electromagnetic spectrum that bear information through air or wires in telecommunications – DNA is a neutral carrier of information, independent of its chemistry and physics... As the Central Dogma ordains and information theory dictates, the DNA program is discrete and digital, and its information is transferred through chemical carriers – but it is not specified by chemical forces. Each unit of biological information is passed on according to a digital program – a biological code – that is transcribed and translated into amino acids" [Gil06]. Information theory yields Crick's central dogma.

Shannon information theory [Sha48] deals with the reduction of possibilities or uncertainty such that the amount of "information" (actually complexity) in a string of symbols is inversely related to the probability of the occurrence of that string. Shannon uncertainty provides only a mathematical measure of improbability, not whether a symbol string is meaningful or significant (no functionality is required).

Shannon information (often referred to as information entropy) quantifies the data contained in a message as the minimum message length necessary to communicate that data. This also represents the absolute limit on the best possible lossless compression for any storage or communication of that data. For example, archiving programs like Zip can create a shorter file than the original by translating the original message into a message with the same Shannon information, but with a more concise representation using a compression algorithm like Lempel-Ziv [Ziv78]. "Lossless" means that when uncompressed, an exact copy of the original data is obtained.

The Shannon information of a message is the minimum bits per symbol multiplied by the number of symbols in the message. The minimum bits per symbol is usually not an integer. For example, English text typically has 0.6-1.5 bits per letter [Sha50], which is considerably below the five (or six for case sensitivity, or eight as it's usually stored) bits per letter to represent that text in a computer.

A long string of repeating characters has 0 information as soon as every character is predictable, e.g. in "junkjunkjunk...," only the first "junk" contributes to the information. A string of random letters in the range 'A'-'P' would have 4.0 bits per letter since there are 2^4 (= 16) choices with no predictability. The Shannon information is a measure of the average content that is unknown (see Appendix D for equations).

In a pair of tossed dice, a sum of 2 or 12 has very high information content, since each die would be known for certain (either 1 or 6). A sum of 7, on the other hand, has low information content, since there is considerable uncertainty of the value of each die. In rolling three dice, sums of 3 and 18 would convey the most information about the dice.

Low Information (less specific): 6 ways to make 7 **High Information:** 1 way to make 2 or 12

A coin toss has one bit of Shannon information since the result of each toss can only be predicted to 0.5 probability. A 2-headed coin, on the other hand, would very quickly degenerate to 0 additional information since one could predict the result with a high degree of certainty. Note that random data, although it has zero functional information, has the maximum possible probabilistic complexity since there is no predictability and therefore the "message" cannot be compressed using a more concise alphabet.

The Shannon information of this book is higher than that of typical English text since subscripts, superscripts, and special characters are used. In an early version of a similar book [Joh09I], the Zip program reduced 5 chapters from 85,683 to 33,278 bytes, so the Shannon information was known to be under 266,224 bits (8 bits per byte). If Zip's compression alphabet is double the theoretical minimum, the estimated Shannon information in those 5 chapters was about 133K bits.

DNA is an extremely stable storage system. In the DNA sequence, there are strong ester bonds joining the nucleotides, with no chemical determination as to which base should be adjacent to another base. Since each position is totally free to take any base, DNA is an ideal information storage system. If a particular base had a tendency to attach next to another particular base, the information-carrying capacity would be reduced (to zero if bonding was determined by "law").

The stability of the DNA helix, in which two strands wrap around each other, is determined by the bases. Two bases, A and G, have large double-ring (purine) structures, while C and T have smaller single-ring (pyrimidine) structures (see Appendix C). The hydrogen bond between two bases in adjacent strands always involves a weak bond between one of each sized molecule, so that bonds are always A-T or C-G between the helix strands. This results in a constant distance between the weakly-bonded bases to produce a very stable helix structure. The double helix structure makes DNA very robust.

One ramification of that robustness is that the information stored is reduced by 50% since one strand is totally redundant, being a base-paired complement of the other strand (if one strand is known, so is the other). Although human DNA contains about 12 Gbits of data in its 6 billion bases, only 6 Gbits are information, since the other 6 Gbits in one strand are totally determined by the other strand. Either strand could be considered to be information, but once that choice is made, the other strand adds nothing new, and is therefore not information (although it does add data).

The genetic information system is analogous to a computer's information as determined from information theory, since both are segregated, linear, and digital [Ben73, Cha79]. In bioinformatics (the formal study of the information in life), segregated means each codon is a distinct symbol, linear means these symbols are in a distinct meaningful sequence in the DNA or RNA, and digital means there's no blending of characteristics of symbols and no lowering of fidelity during copying or communicating. As the hype concerning the "Big Switch" to digital TV has indicated – digital is better (it's interesting that life "knew" that from its beginning).

"The genetic information system *operates without regard for the significance or meaning of the message, because it must be capable of handling all genetic messages of all organisms, extinct and living, as well as those not yet evolved... The genetic information system is the software of life and, like the symbols in a computer, it is purely symbolic and independent of its environment"* [Yoc05p7]. *"Wherever you go in the world, whatever animal, plant, bug or blob you look at, if it is alive, it will use the same dictionary and know the same code. All life is one. The genetic code, bar a few tiny local aberrations, mostly for unexplained reasons in the ciliate protozoa, is the same in every creature. We all use exactly the same language"* [Rid99]. *"It seems that the two-pronged fundamental question: 'Why is the genetic code the way it is and how did it come to be?', that was asked over 50 years ago, at the dawn of molecular biology, might remain pertinent even in another 50 years. Our consolation is that we cannot think of a more fundamental problem in biology"* [Koo08].

The genetic code describes (technically, maps via code bijection) the correspondence between each codon triplet and its corresponding amino acid in a semiotic system that characterizes the symbols and their meaning. A code's symbolic alphabet contains the minimum number of symbols for such mapping. A symbol may have multiple components, such as a codon that has 3 bases, each of which is one-of-four, so the total number of symbols in the genetic alphabet is 64 (= 4^3). The protein alphabet has 20 symbols since there are 20 amino acids.

Since life involves information, it must follow the rules of information theory, including that of transferring information. Yockey proves that it is impossible (zero probability) to transfer information from the 20-symbol protein alphabet to the 64-symbol genetic code. *"Since no code exists to transfer information from protein sequences to mRNA, it is impossible for the origin of life to be 'proteins first'"* [Yoc92]. This categorically eliminates the protein first scenarios as the origin of life, since *"it is mathematically impossible, not just unlikely, for information to be transferred from the protein alphabet to the mRNA alphabet"* [Yoc05p23]. "*Scientists cannot get around it by clever chemistry. This restriction prevails in spite of* [what] *the concentration of protein in a 'prebiotic soup' may have been or may be on some 'Earth-like' planet elsewhere in the universe"* [Yoc05p182]. This is the information theory basis for Crick's "Central Dogma" that states that information transfer is only from DNA to protein, and never the reverse.

Yockey shows that since the genetic code had to be present from the very beginning of life, *"the origin of life, like the origin of the universe*[,] *is unknowable. But once life has appeared, Shannon's Channel Capacity Theorem ... assures us that the genetic messages will not fade away"* [Yoc05p181].

Yockey has shown the impossibility of transferring information from an alphabet of fewer symbols to one with more basic symbols. As proved in Appendix D, whatever the source of life (which is scientifically unknowable), the alphabet involved with the origin of life, by the necessary conditions of information theory, had to be at least as symbolically complex as the current codon alphabet. If intermediate alphabets existed (as some have speculated), each predecessor also would be required to be at least as complex as its successor, or Shannon's Channel Capacity [Sha48] would be exceeded for information transfer between the probability space of alphabets with differing Shannon capacity. Therefore, life's original alphabet must have used a coding system at least as symbolically complex as the current codon alphabet. There has been no feasible natural explanation proposed to produce such an alphabet since chance or physicality cannot produce functional information or a coding system, let alone a system as complex as that in life. See Appendix D for the technical details of the findings just summarized.

This chapter has highlighted how the principles of information theory apply to life. Shannon information theory deals only with the probability of the data, without regard to any functionality. Shannon information theory is important in that it supplies limits on the generation, storage, communication, and translation of data. "Complexity" or "uncertainty," are actually better terms than "information" for describing this probabilistic characteristic that may or may not contain anything meaningful. Random data, for example, has the highest possible complexity, uncertainty, and Shannon information. Information theory also highlights the necessity of codes for reliable data communication. The next chapter will cover the much more important prescriptive information (PI) in more depth. Chapter 7 will summarize how the different types of information work together in life to handle the cybernetic complexity and functionality that is clearly evident, as was highlighted in chapter 4.

6 Prescriptive Programming Information in Life

"Peer-reviewed life-origin literature presupposes that, given enough time, genetic instructions arose via natural events. Thus far, no paper has provided a plausible mechanism for natural-process algorithm-writing... Both the semantics [meaning] *and syntax* [grammar rules] *of codonic language must translate into appropriate semantics and syntax of protein language. That symbolization must then translate into the 'language' of three-dimensional conformation via minimum-free-energy folding* [for a stable protein]. *No combination of the four known forces of physics can account for such conceptual relationships. Symbolism and encryption/decryption are employed. Codons represent functional meaning only when the individual amino acids they prescribe are linked together in a certain order using a different language. Yet the individual amino acids do not directly react physicochemically with each triplet codon. Even after a linear digital sequence is created in a new language, 'meaning' is realized at the destination only upon folding and lock-and-key binding"* [Tre04].

Formal algorithmic prescriptive information (PI) is key to any successful computer program, including the programs within life. A successful algorithm will have "computational halting" – that is, the program will stop when (not if) its problem is solved. Note that some programs, e.g. controllers, are designed to execute forever, so they may be successful even if their "problem" hasn't been totally solved (since more input is possible). PI doesn't just describe, it generates meaning and function, providing a recipe or functional algorithm. A programming "bug" is an example of a slight mis-prescription, which makes the program less-than-ideal, or even worse-than-useless (getting no result may be preferable to a wrong result). The essence of prescription itself is choice contingency. Purposeful intent is required at each successive decision node in order to choose configurable switch-settings (or equivalent) and to steer events toward functional results.

Prescriptive information (PI) *"instructs or directly produces nontrivial function... Prescriptive information either tells us what choices to make, or it is a recordation of wise choices already made"* [Abe09P]. PI involves formal choices at decision points that cannot be generated by randomness or law (necessity). For example, a computer program is formally solved algorithmically before physically implementing it (a program doesn't just appear on a disk without instantiated choices). This chapter will show that the information in life is fundamentally formal, not just physical. Life's information is stored in the physical DNA medium, but the nucleotide sequence of DNA contains PI that is clearly evident.

"For prescription to be realized, the destination of any message must have knowledge of the source's alphabet, rules, and cipher. The destination must also possess the ability to use the cipher" [Abe09P]. A cipher is an algorithm to perform encryption or decryption. For example, if the message "Ju xpslt" were received according to the agreed transmission protocol between sender and receiver, it is a meaningless message without the cipher. If decryption requires replacing a letter with its predecessor, the message "It works" is the result.

PI isn't limited to computer algorithms, but characterizes any step-by-step procedure to accomplish a specific purpose. For example, the instruction sheet to assemble a bicycle is PI. Having all the parts is insufficient for generating a functional result. Envision how long it would take to obtain anything useful if a blindfolded person randomly chose parts from a pile, putting them together. Envision how long it would take if the person had no idea what was being constructed and had never seen or envisioned a bicycle (construction without purpose). On the other hand, if that person were furnished the parts, tools, and specific instructions when needed, a functional result is quite likely.

When it comes to instructions, recipes, messages, and cybernetic programs, grammar rules (syntax) cannot be isolated entirely from message meaning (semantics) or message function (pragmatics). Syntax without meaning also lacks function. Thus PI requires all three categories of symbols and their meaning (**semiotics**) to communicate shared meaning and function between source and destination. *"Biosemiotics can be defined as the science of signs in living systems"* [Kul07].

Cybernetics is the interdisciplinary study of control systems with feedback. Without specifying the origin of cybernetic systems, *"Cybernetics is the study of systems and processes that interact with themselves and produce themselves from themselves"* [Kau07]. It may be tempting to view the physical semiotic systems of life as purely physical. When other cybernetic and artificial life systems are examined, it's clear that they function only because of formal controls instantiated by hardware and software physicality. *"But when it comes to life's syntax, semantics, and pragmatics, we fanatically insist for metaphysical reasons that the system is purely physical. No empirical, rational, or prediction-fulfillment support exists for this dogma"* [Abe09P].

"Metaphysically disallowing formalism in one's model of reality precludes not only redundancy coding, it precludes semiosis. A purely physical semiotic system cannot exist or function as a messaging system. Representationalism [symbols represent, e.g. – 'cat' is a representing word] *requires both combinatorial uncertainty* [other letters are possible] *and freedom of deliberate selection... Formalism alone can send and interpret linear digital messages. This remains true even when a material symbol system* [MMS] *with physical symbol vehicles is used by formalism. Polynucleotide genes are such an MSS*" [Abe09G]. It is impossible to send a meaningful message non-symbolically. It has been recognized that *"the cell is a true semiotic system, and that the genetic ... codes are experimental realities"* [Bar08S].

Formalism is an extremely important reality to recognize in life. Because of metaphysical beliefs, many scientists recognize only chance and necessity, and dismiss out-of-hand any non-material reality. This is inconsistent with their accepting the reality of the formal physics properties, e.g. gravity with its formal equations, even though it's unknown why (purpose) or how (mechanism) it works (but it does -- every time). Note that none of the papers used to establish formalism as a reality invoke anything "supernatural" as its cause (see Appendix G). For example, in peer-reviewed papers [Bar08S, Bar08B], Barbieri considers how formal "code-makers" required for biosemiosis may have arisen by purely natural processes. The Salzburg conference on "Natural Genetic Engineering and Natural Genome Editing" investigated how to integrate these formal concepts into understanding evolution [Wit09].

"Biological information is not a substance... biological information is not identical to genes or to DNA (any more than the words on this page are identical to the printers ink visible to the eye of the reader). Information, whether biological or cultural, is not a part of the world of substance" [Hof05]. It is critical to understand the difference between something physical and its symbolic representation. Symbols can be instantiated in any number of different media without changing the meaning of those symbols. For example, a word can appear in printed, hand-written, projected, electronically transmitted, encrypted, etc. form without changing its meaning.

"Genetic algorithms instruct sophisticated biological organization. Three qualitative kinds of sequence complexity exist: random (RSC), ordered (OSC), and functional (FSC). FSC alone provides algorithmic instruction... Law-like cause-and-effect determinism produces highly

compressible order. Such forced ordering precludes both information retention and freedom of selection so critical to algorithmic programming and control. Functional Sequence Complexity requires this added programming dimension of uncoerced selection at successive decision nodes in the string. Shannon information theory measures the relative degrees of RSC and OSC. Shannon information theory cannot measure FSC. FSC is invariably associated with all forms of complex biofunction, including biochemical pathways, cycles, positive and negative feedback regulation, and homeostatic metabolism. The algorithmic programming of FSC, not merely its aperiodicity, accounts for biological organization. No empirical evidence exists of either RSC or OSC ever having produced a single instance of sophisticated biological organization. Organization invariably manifests FSC rather than successive random events (RSC) or low-informational self-ordering phenomena (OSC)" [Abe05]. For example, a fractal produces amazingly intricate visual patterns, but is an example of OSC whose recursive algorithm contains little information.

The rise of PI presumably occurred early in the evolutionary history of life which had several organisms that depended upon nearly 3,000 highly coordinated genes in cybernetic systems. Genes are linear, digital, cybernetic sequences. They are meaningful, pragmatic (functional), physically instantiated recipes. Bioinformation has been selected algorithmically at the sequence level to instruct eventual three-dimensional shape. The shape is specific for a certain structural, catalytic, or regulatory function, with all functions integrated into a symphony of metabolic functions. The sequence of DNA nucleotides ultimately produces the protein's shape, but using a very indirect mechanism that includes multiple conversions of digital information.

It has long been recognized that the genotype (gene) is a generative algorithm – *"a carefully spelled out and foolproof recipe for producing a living organism... the algorithm must be written in some abstract language... a language must have rules"* [Ede66].

It is important to differentiate natural law constraints from controls. Investigators can choose initial conditions for the starting point of their experimentation, but those "chosen constraints" can be considered controls because they were deliberately selected to steer events toward desired results. Choice contingency alone, not the constraints themselves, achieves non trivial integration, organization, and function. *"Controls do not arise from the categories of chance contingency and necessity addressed by thermodynamics, kinetics and physics in general. Physics can address constraints. Physics cannot address bona fide controls without acknowledging the reality of non naturalistic engineering. Life is wholly dependent upon tight regulation and controls. For this reason, physics and chemistry alone cannot adequately address and explain life any more than physics and chemistry alone can explain engineering"* [Abe10]. Formal choice contingency can control physicality.

"What kind of information produces function? In computer science, we call it a 'program.' Another name for computer software is an 'algorithm.' No man-made program comes close to the technical brilliance of even Mycoplasmal genetic algorithms. Mycoplasmas are the simplest known organism with the smallest known genome, to date. How was its genome and other living organisms' genomes programmed?" [Abe05] This is probably the biggest unanswered question in biology since algorithms are unknown except for choice contingency.

The bits in a computer program measure only the number of binary choice opportunities in the algorithm. Bits do not prescribe which choices are made, but instead prescribe the actions resulting when specific data are processed. A computer program has a sequences of integrated specific decision-node choice-commitments that we expect to be useful when the program is purchased.

The goal of genome projects has been the enumeration of the sequence of the particular choices in the DNAs. Both computer computations and the end-products of biochemical pathways require non-random algorithmic processes or procedures that produce the needed results Those algorithms are never "self-ordered" by redundant cause-and-effect necessity. Each successive nucleotide in DNA or RNA is a quaternary (base-4) "switch setting," produced by uncoerced selection. This cybernetic aspect of life processes is directly analogous to that of computer programming, and requires considerably more emphasis and attention than it typically receives.

"What sense can we make, then, of the PI found in nature and particularly in any theorized primordial biosemiosis? Random coursing through a succession of bifurcation [decision forking] *points has never been observed to lead to prescription of function, computational halting, sophisticated circuitry, or system organization. The self-ordering events described by chaos theory cannot generate conceptual formal organization. Semiosis, cybernetics, and formal organization all require deliberate programming decisions, not just self-ordering physicodynamic redundancy. Self-ordering phenomena are low-informational, highly*

redundant, unimaginative, and usually destructive of organization (e.g., tornadoes and hurricanes). No prediction fulfillments have been realized of spontaneous natural events producing formal algorithmic optimization. No empirical support or rational plausibility exists for blindly believing in a relentless natural-process assent up the foothills of a rugged fitness landscape toward mountain peaks of formal functionality. Investigator involvement creates this illusion usually through the hidden artificial steering of experimental iterations" [Abe09P].

Life is orchestrated by the prescriptive information (instruction) content and the inherent systems it has from its inherited genome. Cells also are dependent upon their current environment, especially for energy, whose transducing mechanisms are instructed by its genome. Life is an integration of many algorithmic processes that give rise to biofunction and overall life-maintaining metabolism. Each symbolic choice is critical to the determination of eventual function, so that bioinformation is more than just physical order, complexity, or probabilistic uncertainty, but is an instructive, orchestrational recipe. Unlike Shannon complexity, PI is concerned with functionality, not the degree of compressibility.

Take note anyone that wants to earn an "easy" $1 million: *"The Origin-of-Life Prize®... will be awarded for proposing a highly plausible mechanism for the spontaneous rise of genetic instructions in nature sufficient to give rise to life"* [OOLprize]. *"The Origin-of-Life Foundation, Inc. is a science and education foundation encouraging the pursuit of natural-process explanations and mechanisms within nature."* Since no theory of genetic information is complete without a model of mechanism for the source of such prescriptive information within Nature, *"all submissions must address the source of the prescriptive information through non-supernaturalistic natural processes. Which of the four known forces of physics, or what combination of these forces, produced prescriptive, functional information, and how? What is the empirical evidence for this kind of prescriptive information (instruction) spontaneously arising within Nature?"* [OOLprize].

In all known life, genomes manifest functional coded messages, using a sign system, to distant sites through an information channel to energy-consuming decoding receivers in ribosomes. Languages are translated from one symbolic, indirect representation of the message from one alphabet into another, specifically from the codon language into the language of the physical amino acid sequence in the end-product. Explanation is required for the instructions capable of causing and affecting the multitude of individual manufacturing processes and coordinating of all of those diverse processes toward the apparent "purpose" of being alive. What natural mechanism(s) caused *"the initial writing of this prescriptive information by nature, not just the modification of existing genetic instruction through mutation"* [OOLprize].

The after-the-fact modification of an existing genome is different from the production of the genome from non-living components. The Genetic Selection Principle *"distinguishes selection of existing function (natural selection) from selection for potential function (formal selection at decision nodes, logic gates and configurable switch-settings)"* [Abe09G]. Undirected nature has no "purpose," and the source of functional algorithmic PI that "will have purpose" at some future time has yet to be explained. If RNA-first is true, *"the real issue of life origin lies in answering how the initial single positive strands of RNA instructions got sequenced so as to prescribe microRNA regulation, amino acid sequencing and eventual folding function. No new information is generated in base-pairing replications... Base-pairing is purely physicodynamic, and quite secondary to the already-programmed, formal, linear digital instructions of the single positive strand... Prior to an algorithm having computational function, no basis exists for selection in nature... How did any computational program arise in nature? Computation is formal, not physical. Natural selection cannot generate formalisms. It can only prefer the results of formal computations—already living organisms... No organism exists without hundreds of cooperating formal algorithms all organized into one holistic scheme. The more computational steps that are required to achieve integrated halting, the harder it becomes for an inanimate environment to explain optimization"* [Abe09G].

"How did inanimate nature give rise to an algorithmically organized, semiotic and cybernetic life? Both the practice of physics and life itself require traversing not only an epistemic cut, but a Cybernetic Cut – a fundamental dichotomy of reality" [Abe08]. The "epistemic cut" describes the unavoidable gulf between an object and knowledge about that physical object (between description and the thing being described) [Pat67]. *"The Cybernetic Cut must be crossed to program computational halting into any form of physical hardware. To prescribe, instruct or program formal utility is to traverse The Cybernetic Cut... The bifurcation points found in the simplest binary system of choice contingency are bona fide decision nodes. Crossing the Cybernetic Cut requires the*

ability to purposefully steer through successive bifurcation points down a path toward a desired goal... Bifurcation points, in the -absence of the intentional choice that would convert them to true decision nodes, consistently fail to generate sophisticated utility. In symbol systems, the randomization of symbols and denial of intentional symbol selection quickly leads to the loss of even rudimentary meaning and function... What exactly is the missing ingredient that renders life unique from inanimate physics and chemistry? The answer lies in the fact that life, unlike inanimacy, crosses the Cybernetic Cut" [Abe08].

There is *"a critically important distinction between order and the informed functional organization that characterizes living systems... our paradigm is cybernetic"* [Cor00]. True organization requires crossing the Cybernetic Cut in order to produce a cybernetic system. The resulting system is functional using physical components, but physicality has no mechanism to produce a cybernetic system requiring formal organization.

The RNA/DNA sequence structure is formed prior to protein folding of the PI's result. The fixed linear digital highly informational nucleotide sequence must be prescribed semiotically before the weak hydrogen-bond folding of the secondarily protein structure occurs. The Genetic Selection Principle, *"states that selection must operate at the genetic level, not just at the phenotypic* [structural] *level, to explain the origin of genetic prescription of structural and regulatory biological function. This is the level of configurable switch-settings (nucleotide selection). Selection must first occur at each decision node in the syntactical string. Initial programming function cannot be achieved by chance plus after-the-fact selection of the already-existing fittest programs (phenotypes). Evolution is nothing more than differential survival and reproduction of already-existing fittest phenotypes. The computational programming proficiency that produced each and every phenotype* [observable characteristic] *must first be explained... Thus far, no natural process explanation has been published for selection at the decision-node, configurable-switch, nucleotide- selection level"* [Abe08].

No feasible source of life's prescriptive algorithms has been proposed to date. The speculations proposed thus far are"dreams," based on the belief that physicality is the only reality, and therefore "must have" produced life's cybernetic complexity. Information science, by contrast, recognizes non-material formalism as the only reality capable of producing PI.

7 Combining Life's Information Types

To illustrate how the different types of information are combined in a relatively simple multiple-computer system, consider what's involved when making a new remote control unit work with a DVD player. Both the DVD player and remote controller have built-in special-purpose computers, with communication enabled from the remote to the player. The Philips SRU2103/27 control has the prescriptive information (probably entered into a wordprocessor as functional information) for "Direct Code Entry."

1. Press and hold the SETUP button until the red indicator stays on, then release the SETUP button.
2. Press and release the desired mode button (TV, VCR, etc.). The red indicator blinks, then stays lit.
3. Enter the 4 digit code from the code tables (on separate sheet). Note: after a valid code entry, the red indicator turns off. For an invalid code, the red indicator flashes.
4. With your device on, press CHANNEL UP. If the device responds, no further programming is required.

Symbols are used to communicate the meaning of the controller's buttons and the coded information from the 4-digit table to make the controller's communication protocol match that of the player. Set-up requires following the human-controller protocol specified. Once remote-player protocols have been established, pressing a button will result in the controller's computer algorithmically decoding the pressed button. That computer will then issue the needed output commands to create a series of infrared light pulses transduced for communication with the player's receiver. The player receives the transduced signals and its computer algorithmically decodes them to produce the needed electrical controls for the player's mechanism.

The same types of formal symbol use, encoding/decoding, communication protocols, etc. are used in biological systems. Obviously, the hardware involved is different, as are the algorithms and coding. But it is important to realize that the information-handling components in each cell are as real as those involving electronic computers.

Computer/electronic engineers have designed the hardware and interpreting microcode to read the computer's native language instructions (non-physical machine language) and execute the operations specified by each instruction. A higher-level computer language (such as

BASIC, C, or FORTRAN) may be used to translate a desired algorithm into the computer's native language. A computer's operating system (OS), such as Linux or Windows, is a set of programs that allows applications programs to execute on the computer's hardware, allowing access to storage and input/output devices. To the computer hardware, an OS is just another program, unless it's built into the hardware. To an application program, typically only the OS (not the hardware) is visible. In the history of computers, there has never been an instance of functional hardware or software arising by undirected processes (although some non-functional student-written programs have had the appearance of being the result of random trials).

Advances in molecular biology, biochemistry, and bioinformatics (the study of the information of life) have revolutionized our understanding of the cell's miniature world. Cells possess the ability to store, edit, and transmit information and to regulate metabolic and other processes using that information. Whereas cells were once thought of as simple *"homogeneous globules of plasm,"* [Hae04] by biologists of Ernst Haeckel's time 100 years ago, today a cell is *"viewed as a complete computational machine in terms that are akin to a multi-core computer cluster, where there is a centralized memory and instruction set, yet computational tasks are distributed among distinct processing elements"* [D'On10].

Among the "Universal Laws of Information" [Git97], **it is impossible** (zero probability): 1) to set up, store, or transmit information without using a code; 2) to have a code apart from deliberate convention determined by rules (not law); 3) to have information without a sender; and 4) that information can exist apart from a formal source.

From the information perspective, the genetic system is a preexisting operating system of unknown origin that supports the storage and execution of a wide variety of specific genetic programs (the genome applications), each program being stored in DNA. DNA is a storage medium, not a computer, that specifies all information needed to support the growth, metabolism, parts manufacturing, etc. for a specific organism via gene subprograms.

DNA has been compared to a computer's disk drive (Appendix B) [D'On10], which makes sense in a NUMA (non-uniform memory access) model. Early real computers used disk-like drums and other sequential-access main memories. This author has peer-reviewed publications [Joh95, Joh97B, Joh05] describing concepts of distributed sequentially-accessed special-purpose and general memories (analogous to those in life) in heterogeneous (different) multiprocessor systems. In cells, there are many RNAs and micro-proteins, most with unknown functions, which may function as registers and inter-processor communications channels.

Technically, DNA is an example of shared memory in a distributed heterogeneous multiprocessor system with Flynn classification multiple input streams and multiple output streams [Fly72]. For DNA, there are multiple differing enzyme computers simultaneously reading different portions of the DNA genetic code, each producing its own output (for example, via mRNA). Each cell has over 2,000 different enzyme computers that read the shared memory data in DNA, processing that data according to the individual programs, many operating independently (though many operations require multiple cooperating enzymes).

The native language includes a coding system (e.g. – codon-based encryption) whose codes are read by enzyme "computers." Often, smaller sections of RNA from different genes are spliced together to form the mRNA for a particular protein specification. The mRNA output is ultimately to another OS in a ribosome, which has its own program stored in its RNA, where the codes are decrypted. The needed output signals are then transmitted to the tRNA computer (which has been programmed to pick up its associated amino acid via its own program and OS) so that the amino acid specified by the codon is transported to the construction site to be added to the protein being built.

"Due to the abstract character of function and sign systems [semiotics -- symbols and their meaning], *life is not a subsystem of natural laws. This suggests that our reason is limited in respect to solving the problem of the origin of life and that we are left accepting life as an axiom... Life express*[es] *both function and sign systems, which indicates that it is not a subsystem of the universe, since chance and necessity cannot explain sign systems, meaning, purpose, and goals"* [Voi06]. Necessity refers to characteristics determined by "law," such as a released object falling due to gravity, or burning hydrogen in oxygen to produce water.

The coded information system in a cell *"may be compared to a book or to a video or audiotape, with an extra factor coded into it enabling the genetic information, under certain environmental conditions, to read itself and then to execute the information it reads. It resembles, that is, a hypothetical architect's plan of a house, which plan not only contains the information on how to build the house, but which can, when thrown into the garden, build entirely of its own initiative the house all on its own without the need for contractors or any other outside building agents... Thus, it is fair to say that the technology exhibited by the genetic code is orders of magnitude higher than any technology man has, until now, developed. What is its secret? The secret lies in its ability to store and to execute incredible magnitudes of conceptual information in the ultimate molecular miniaturization of the information storage and retrieval system of the nucleotides and their sequences"* [Wil87].

Every case of coded information, where the source is known, invariably requires formalism for its creation. Bill Gates, founder of Microsoft, writes, *"Human DNA is like a computer program but far, far more advanced than any software we've ever created"* [Gat96]. Can you imagine how believable it would be if someone were to suggest that the Windows 7 operating system just arose from physicality?

Functional information principles may be used to quantify traditional information by calculating a value which *"represents the probability that an arbitrary configuration of a system will achieve a specific function to a specified degree"* [Haz07]. Kalinsky [Kal08] evaluated examples using quantified functional information, comparing the probabilities of such information arising by physical and formal causes. The five "watermarks" (words whose amino acid letters are meaningful, including "CRAIGVENTER") of Venter Institute's synthetic M. genitalium genome were found to be 10^{22} times more probable to have a planned (known in this case) as opposed to purely physical source. A 300 amino acid protein was found to be 10^{155} times more probable to have a planned source. The simplest life form with 382 protein-coding genes [Gla06] was found to be $10^{80,000}$ times more probable to have a planned source. Appendix E has technical details.

Unlike Shannon complexity, which is independent of meaning, the functional information assumes the receiver of the information in this book (you, the reader) can appropriately decode the blotches of ink on the white background to gain information based not only on the forms of those blotches, but on knowledge in the receiver's data-base (you know the meaning of English words). For example, there was very little new information in chapter 1 for most scientists (nearly everything was predictable), whereas a non-scientist most likely had a considerable information increase while reading that chapter. If this book were shown to someone not familiar with English, it's likely that little information would be conveyed, despite the book's information content.

This concept is critical since it is important not only to transmit the messages reliably (a Shannon criterion) from the DNA, but the enzymes and ribosomes must already know how to interpret that coded information in order for proteins (including enzymes), RNA, and replicated DNA to be manufactured appropriately (functional criteria). In other words, the data transmitted may have an effective Shannon information content that exceeds the number of raw data bits transmitted, if one includes all of the information already known by the receiver. When "DNA" appears in this text, for example, considerably more than three English letters is being communicated. If a DNA molecule were placed into a pool of amino acids, the information in the DNA would be useless unless the needed receiver enzymes were present with prior information *"necessary to function"* [Sar96].

"Neither order nor complexity is the determinant of algorithmic function... This is one of most poorly understood realities in life-origin science. Selection alone produces functionality. Without selection, evolution would be impossible... A 'cybernetic program' presupposes a cybernetic context in which it operates. One has to have an operating system of 'rules' before one can have an application software. And of course one must have a hardware system too. All of these components only come into existence through 'choice contingency,' not through 'chance contingency' or law. One of many problems with metaphysical materialism is that it acknowledges only two subsets of reality: chance and necessity. Neither can write operating system rules or application software. Neither can generate hardware or any other kind of sophisticated machinery, including molecular machines (the most sophisticated machinery known)" [Abe07].

"The reductionist approach has been to regard information as arising out of matter and energy. Coded information systems such as DNA are regarded as accidental in terms of the origin of life and that these then led to the evolution of all life forms as a process of increasing complexity by natural selection operating on mutations on these first forms of life" [McI09]. In the last 10 years, at least 20 different natural information codes were discovered in life, each operating according to arbitrary conventions (not determined by law or physicality). Examples include proteins address codes [Ber08B], acetylation codes [Kni06], RNA codes [Fai07], metabolic codes [Bru07], cytoskeleton codes [Gim08], histone codes [Jen01], and alternative splicing codes [Bar10].

Much of the functional information of life is involved with

cybernetic control, in which life's components regulate and control where needed. *"All the equations of physics taken together cannot describe, much less explain, living systems. Indeed, the laws of physics do not even contain any hints regarding cybernetic processes or feedback control. Thus, the term 'dissipative structures' does not adequately describe the 'informed', purposive organization of living systems. It is comparable to characterizing jet engines -- which are painstakingly designed and manufactured with extremely precise dimensional properties and tolerances -- as dissipative structures. They are neither self-designed nor are their dissipative properties among their most salient features"* [Cor05]. Many RNAs are transcribed from DNA to be regulators (necessary for "control"), including of mRNA. *"In addition to revealing the surprising diversity of post-transcriptional events that regulate mRNAs, our work also points to new roles for a family of proteins that mediate RNA interference"* [Han10].

Researchers have been able to identify a mechanism that turns a maturing embryo's developmental genes off and on. *"By understanding how development unfolds, we can better control this process, which includes cell proliferation and organ development"* [Yeh10]. Scientists have found that many proteins are multi-functional from the transcription, metabolism, and manufacturing levels. *"At all three levels, we found M. pneumoniae was more complex than we expected"* [Ser09].

"The complexity of biology has seemed to grow by orders of magnitude... the signaling information in cells is organized through networks of information rather than simple discrete pathways" [Hay10]. Biocomplexity professor Stuart Kauffman points out,*"the genetic regulatory network in humans has some 23,000 genes, among which are at least 2,040 transcription factor genes. These TFs regulate one another's transcriptional activity and those of genes that are regulated but not regulating. Work on yeast gene networks shows that they appear to be one large interconnected network... This genetic regulatory network is a non-linear dynamical system of high complexity. Modeling genes as binary, on, off, devices and studying large 'random Boolean networks' has shown that these networks, and piecewise linear networks, and linear ordinary differential equation networks all show the same generic behaviors,"* all being complex data processing implementations [Maz10p223].

Scientists are investigating *"the organization of information in genomes and the functional roles that non-protein coding RNAs play in the life of the cell. The most significant challenges can be summarized by two points: a) each cell makes hundreds of thousands of different RNAs and a large percent of these are cleaved into shorter functional RNAs demonstrating that each region of the genome is likely to be multifunctional and b) the identification of the functional regions of a genome is difficult because not only are there many of them but because the functional RNAs can be created by taking sequences that are not near each other in the genome and joining them together in an RNA molecule. The order of these sequences that are joined together need not be sequential. The central mystery is what controls the temporal and coordinated expression of these RNAs"* [Gin10].

"It is very difficult to wrap your head around how big the genome is and how complicated... It's very confusing and intimidating... The coding parts of genes come in pieces, like beads on a string, and by splicing out different beads, or exons, after RNA copies are made, a single gene can potentially code for tens of thousands of different proteins, although the average is about five... It's the way in which genes are switched on and off, though, that has turned out to be really mind-boggling, with layer after layer of complexity emerging" [LeP10].

For functional communication (including controls) to occur, both sender and receiver of each communication step must know the communication protocol and how to handle the message. In each cell, there are multiple OSs, multiple programming languages, encoding/decoding hardware and software, specialized communications systems, error detection and correction mechanisms, specialized input/output channels for organelle control and feedback, and a variety of specialized "devices" to accomplish the tasks of life. The author can attest that these concepts are not trivial since many were fundamental to his second Ph.D. thesis research [Joh97T].

The challenge for a purely physical origin of such a cybernetic complex interacting computer system is the need to demonstrate that the rules, laws, and theories that govern electronic computing systems and information don't apply to the even more complex digital information systems that are in living organisms. Laws of chemistry and physics, which follow exact statistical, thermodynamic, and spacial laws, are totally inadequate for generating complex functional information or those systems that process that information using prescriptive algorithmic information. Unfortunately, most people investigating origins are unfamiliar with the immensity of the problems, and believe that time, chance, and natural selection can accomplish almost anything.

It can correctly be asserted that based on known science, some things are impossible. Based on the second law of thermodynamics (entropy/disorganization always increases), a perpetual motion machine is impossible. One needn't prove that each machine that could be conceived is impossible. If it is asserted that such a machine is possible, the one making such an assertion would need to first show that known science is wrong concerning increasing entropy before the assertion is given consideration as science. All the known laws, theorems, and principles of information science indicate that codes, complex functional information, and prescriptive algorithmic information cannot arise from physicality. Note that non-physical formality can be viewed as a completely natural reality [Sha10, Bar08B, Bar08S, Abe09P]. Therefore, based on currently accepted information science, but realizing that science is always subject to change, it seems impossible for life to have arisen purely from physicality.

Those who insist on purely physical causes of life are thus in an untenable position when it comes to known science. Not only can they not prove that it's possible (non-zero probability) for life to come about by the physical interactions of nature, but the information content of life precludes that possibility. *"There is no known law of nature, no known process*[,] *and no known sequence of events which can cause information to originate by itself in matter*" [Git97p107].

Note that one cannot use the information of life to "prove" that information can arise purely from physicality, as that would simply be a tautology based on the assumption of life from physicality. In examining any complex functional information where the source of the information is known, it invariably (no known exceptions) resulted from a formal source, as opposed to chance and/or necessity.

8 Programming Increasing Complexity in Life

"We have seen that living things are too improbable and too beautifully 'designed' to have come into existence by chance. How, then, did they come into existence? The answer, Darwin's answer, is by gradual, step-by-step transformations from simple beginnings, from primordial entities sufficiently simple to have come into existence by chance. Each successive change in the gradual evolutionary process was simple enough, relative to its predecessor, to have arisen by chance. But the whole sequence of cumulative steps constitutes anything but a chance process, when you consider the complexity of the final end-product relative to the original starting point. The cumulative process is directed by nonrandom survival" [Daw96Bp43]. Darwin wrote, *"If it could be demonstrated that any complex organ existed which could not possibly have been formed by numerous, successive, slight modifications, my theory would absolutely break down"* [Dar98p154].

"The Darwinian theory is in principle capable of explaining life. No other theory that has ever been suggested is in principle capable of explaining life" [Daw96Bp288]. This chapter highlights many of the observations that Darwinism has failed to explain. The adequacy of Darwinism as an explanation for life's diversity is in need of closer investigation.

Evolution is this chapter's topic, emphasizing the change in the informational content, especially the increase in information moving up life's biological tree. Computer simulations and evolutionary algorithms will provide reality checks. *"Biological evolution, simply put, is descent with modification. This definition encompasses small-scale evolution (changes in gene frequency in a population from one generation to the next) and large-scale evolution (the descent of different species from a common ancestor over many generations)"* [EVsite].

Microevolution, small adaptive changes that are heritable, is accepted as verifiable fact by virtually all scientists. Macroevolution, the scenario that all life originated by undirected natural processes from an original organism, is believed by most scientists, but doubted by many scientists, including over 800 who have signed the statement: *"We are skeptical of claims for the ability of random mutation and natural selection to account for the complexity of life. Careful examination of the evidence for Darwinian theory should be encouraged"* [Dis-web]. Over 300 physicians have signed the statement: *"As medical doctors we are skeptical of the claims for the ability of random mutation and natural selection to account for the origination and complexity of life and we therefore dissent from Darwinian macroevolution as a viable theory. This does not imply the endorsement of any alternative theory"* [PSSI].

"A persistent debate in evolutionary biology is one over the continuity of microevolution and macroevolution – whether macro-evolutionary trends are governed by the principles of microevolution" [Sim02]. Many extend evolution back to the prebiotic era, so that abiogenesis, life from non-life, (covered in chapter 4 and Appendix C) is also included in chemical (as opposed to biological) evolution.

This chapter will examine what is known about DNA changes through time and the information requirements for the development of new morphology (structures), especially irreducibly complex structures. That DNA can mutate to modify information is fact. That such mutation can produce a net gain in information has yet to be demonstrated, and needs reconciliation with known information science principles. The Cambrian explosion and the biological tree will be examined for their increasing information content. Remember that any biological structure is the result of proteins, each having a prescriptive algorithm required for its formation, which will be highlighted in this chapter. A major focus will be examining beliefs that are purported to be science concerning Darwinian evolution to ascertain their scientific validity. Keep in mind that the vast majority of science is still directed toward supporting Darwinism, and this chapter currently presents "the minority" scientific evidence, which hopefully will encourage a more critical analysis of all findings (how strongly are conclusions supported by data?). Because of many recent findings, with many scientists now questioning the validity of neo-Darwinism as evolution's mechanism, the next edition of this book may relegate more of neo-Darwinism to Appendix F, as a historical footnote. *"A wave of scientists now questions natural selection's role though fewer will publicly admit it"* [Maz10p20].

Only genetic changes can affect evolution, as those changes can be inherited by offspring. Note that DNA changes not affecting reproduction don't affect evolution. For example, when getting a dental x-ray, a lead apron covers the reproductive organs. It's not that those organs are more susceptible to mutation caused by radiation, but that DNA changes there could be inherited by progeny, causing health problems for future generations. Mutations in DNA are normally considered random, and usually occur during cell replication.

Approximately one in 100,000 copied base-pairs has an error [Dar86], but the proof-reading by enzymes reduces point mutation error rate to between 10^{-11} to 10^{-9}. Mutations of genome segments can include duplication, inversion, transposition, insertion, or deletion [Spe97p40]. After a gene mutates, it is duplicated (assuming the mutation doesn't prevent duplication, as usually happens) with the same fidelity as any gene. If a codon is mutated, it may create a different amino acid in the protein it encodes, unless it mutates to a redundant codon for the same amino acid (a built-in error tolerance).

If a mutated nucleotide is in a non-coding section, the effect may be difficult to determine, as those areas are multi-functional. Sometimes a mutation affects a regulatory region at the start of other genes thereby turning them on or off (with potentially many changes).

Neo-Darwinian evolution is described by Dawkins [Daw96C] as random (chance) mutation followed by selection, with selection favoring no mutations. *"Mutation is not an increase in true information content, rather the reverse, for mutation, in the Shannon analogy, contributes to increasing the prior uncertainty. But now we come to natural selection, which reduces the 'prior uncertainty' and therefore, in Shannon's sense, contributes information to the gene pool. In every generation, natural selection removes the less successful genes from the gene pool, so the remaining gene pool is a narrower subset... what is the information about? It is about how to survive"* [Daw08I].

Although some organisms have been known to obtain genetic material from other organisms [Ten00, Szo06, Sha10], most single-cell organisms (early organisms are presumed to be a single cells) reproduce by asexual cell division, with no interaction genetically with any other organism (with no "gene pool" information). A mutation in an asexual organism produces an information loss, as acknowledged by Dawkins. The offspring of that mutated organism would pass on the lowered information to future generations. Each mutation in successive generations would produce additional information losses (which may cause a genetic line to die out). Each mutation produces an information loss, but those losses are supposedly somehow overcome so that cumulative losses produce the gains in information required to evolve to more complex organisms. This seems like the shop-keeper who lost a little on each sale, but made up for it in volume.

Computer simulations and genetic algorithms [Bar54, Bar57] are believed by many to be evidence for the viability of neo-Darwinism.

For example, Dawkins [Daw88] randomly "mutated" by random changes in 28 positions of letters and spaces:

`"WDLTMNLT DTJBKWIRZREZLMQCO P"`

to produce on the 43rd try:

`"METHINKS IT IS LIKE A WEASEL"`

Dawkins' program starts with a nonsense string and *"duplicates it repeatedly, but with a certain chance of random error – 'mutation' – in the copying. The computer examines the mutant nonsense phrases, the 'progeny' of the original phrase, and chooses the one which, however slightly, most resembles the target phrase, METHINKS IT IS LIKE A WEASEL"* [Daw88].

He knew the goal in advance and apparently kept mutations only if they became closer to that goal. Dawkins has not made his algorithm or source code available to the public (he is certainly invited to do so, as any scientist should, so that findings may be tested and verified), making what the program actually does difficult to ascertain. A partitioned search, mutating all unmatched positions simultaneously at each try, stopping further mutations at any position that becomes correct, would produce a quick solution.. With this scenario, the probability of success is high: $1-(26/27)^{28}=0.65$, for at least one desirable mutation on the first try, falling to no lower than 0.038 if only one position is incorrect (known mutation rates are under 10^{-8} [Smi89]). If one allows only unmatched (u) positions in each (of n) offspring to mutate, the probability for success on each trial is $1-(26/27)^{un}$. With an unlimited number of children, it is possible to make a new correct position nearly certain each generation, leading to a solution in about 28 generations.

Dawkins probably used an unspecified mutation rate and an unspecified proximity search with an unspecified number of offspring to determine which progeny to retain. Many different search scenarios were evaluated [Ewe10] using the "best" guesses as to what Dawkins' algorithm might have been. Evolutionary algorithms out-performed a blind search, but were all relatively inefficient at extracting information. The result is that Dawkins' exercise demonstrated nothing related to any understanding of evolution. It did show that a designed program with a specified target would fairly quickly produce the desired result.

Ludwig [Lud93] sponsored an "Artificial Life" contest to find the shortest self-replicating program, with the winning program having 101 bytes. The probability of this program arising by chance is 256^{-101} or 10^{-243}. If 10^8 computers each make 10^7 trials/sec for 3×10^{22} trials/year, a solution becomes probable after 10^{220} years. If a suitable program were half as large, "only" 10^{99} years of processing would be necessary to make probable a self-replicating program by chance. It should be noted that any prescriptive program still requires an operational platform on which to execute. These programs, like all computer programs, were designed and were executed on designed platforms. Information, not random data, caused solutions.

In an electronic computer, *"every silicon chip holds as many as 700 layers of implanted chemicals in patterns defined with nanometer precision and then is integrated with scores of other chips by an elaborately patterned architecture of wires and switches all governed by layers of software programming written by human beings. Equally planned and programmed are all the computers running the models of evolution and 'artificial life' that are central to neo-Darwinian research. Everywhere on the apparatus and in the 'genetic algorithms' appear the scientist's fingerprints: the 'fitness functions' and 'target sequences.' These algorithms prove what they aim to refute: the need for intelligence and teleology* [targets] *in any creative process"* [Gil06].

Avida [Avida] is one of the most widely used [Ada02, Ofr03] artificial life programs. The Avida system's concepts are similar to the Tierra program [Ray92], with a population of self-reproducing strings subjected to random mutations on a machine that can perform any calculation that any other programmable computer can perform (Turing-complete). The virtual computing organisms (NOT, OR, AND, ADD, etc.) can together, theoretically, do any computation using a population of logic and math function organisms. The virtual environment of Avida is initially seeded with a self-replicating human-designed program. This program and its descendants are subjected to random mutations which change instructions within their memory to produce unfavorable, neutral, and favorable mutations.

"Mutations are qualified in a strictly Darwinian sense; any mutation which results in an increased ability to reproduce in the given environment is considered favorable. While it is clear that the vast majority of mutations will be unfavorable – typically causing the creature to fail to reproduce entirely – or else neutral, those few that are favorable will cause organisms to reproduce more effectively and thus thrive in the environment. Over time, organisms which are better suited to the environment are generated that are derived from the initial (ancestor) creature" [Len03].

"Avida makes it possible to watch the random mutation and natural selection of digital organisms unfold over millions of generations" [Zim05]. Avida simulates how random mutations and natural selection produce *"design without a designer"* [Pen08]. The initial 50 instructions contain 15 self-replication instructions that are not subject to mutations. This is an interesting restriction for Darwinian simulation (restricting the DNA for replicating from mutating), but necessary for the program to work at all.

Random mutations that increase genome length or generate useful math or logic functions (often using already evolved functions) on random inputs are rewarded with more CPU time (inheritable). *"While avida is clearly a genetic algorithm (GA) variation (to which nearly all evolutionary systems with a genetic coding can be reduced), the presence of a computationally (Turing) complete genetic basis differentiates it from traditional genetic algorithms. In addition, ... co-evolutionary pressure classifies avida as an auto-adaptive system, as opposed to standard genetic algorithms (or adaptive) systems, in which the creatures have no interaction with each other. Finally, avida is an evolutionary system that is easy to study quantitatively yet maintains the hallmark complexity of living systems"* [AvidaMan]. Reread chapter 4 if you believe that such a computerized system really maintains the hallmark complexity of living systems.

Avida uses an unrealistically small genome, an unrealistically high mutation rate, unrealistic protection of replication instructions, and unrealistic energy rewards. It allows for arbitrary experimenter-specified selective advantages. *"Neglect of key factors or unrealistic parameter settings permit conclusions to be claimed which merely reflect what the decision maker intended a priori... the computer experiments reported using the Avida framework so far have not demonstrated that neo-Darwinian processes could have produced the necessary coding information to produce the hundreds of molecular machines found in natural cells"* [Tru04].

When more realistic fitness values were used, the logic functions found by chance mutations were removed from the genome, so that *"at the end of hundreds of thousands of generations (when the trials were terminated) no logic functions at all were present"* [Tru04]. Although Avida has been cited [Len03] as evidence against irreducible complexity (a later topic), it is significant that when only the most complex function (EQU) was rewarded, it never appeared even in extremely long simulations [Len03]. EQU could only form from simpler functions, and when those are assigned no advantage, the complex function doesn't form. In short, Avida can be used to validate neo-Darwinism only if neo-Darwinism is taken as true (a programmed tautology).

Artificial life programs have demonstrated the need for algorithmic fitness functions, prescriptive information that determines what is "better" (more fit), as well as targets (purpose – where should this lead?).

Neo-Darwinism breaks down the improbability of macroevolution into small changes that allow scaling up the backside of *"Mount Improbable ... inch by million-year inch"* [Daw96Cp77]. In this view, the informational changes can accumulate in the DNA until a positive gene is formed (at that time, only genes that had physical manifestations were thought to contribute to a selective advantage), at which point that gene will be selected for continuation in the new organism's DNA. The non-coding part of DNA became known as "Junk DNA" which is *"the remains of nature's experiments which failed"* [Ohn72].

Dawkins popularized the idea that any DNA not actively trying to get to the next generation would slowly decay away through mutation and that genes are the basis of evolutionary selection [Daw76]. Sagan writes concerning junk DNA, *"some, maybe even most, of the genetic instructions must be redundancies, stutters, and untranscribable nonsense. Again we glimpse deep imperfections at the heart of life"* [Sag92]. Non-coding sections of DNA were seen as the result of mutations that haven't yet resulted in formation of useful genes so that they would provide a selective advantage. This theme was echoed in authoritative textbooks also. *"Introns have accumulated mutations rapidly during evolution, and it is often possible to alter most of an intron's nucleotide sequence without greatly affecting gene function. This has led to the suggestion that intron sequences have no function at all and are largely genetic 'junk'"* [Alb94]. *"Much repetitive DNA serves no useful purpose whatever for its host. Rather, it is selfish or junk DNA, a molecular parasite that, over many generations, has disseminated itself throughout the genome"* [Voe95].

In 1998, restated in 2004, Dawkins wrote, *"there's lots more DNA that doesn't even deserve the name pseudogene. It, too, is derived by duplication, but not duplication of functional genes. It consists of multiple copies of junk, 'tandem repeats', and other nonsense which may be useful for forensic detectives but which doesn't seem to be used in the body itself. Once again, creationists might spend some earnest time speculating on why the Creator should bother to litter genomes with untranslated pseudogenes and junk tandem repeat DNA"* [Daw98, Daw04]. Biology professor PZ Myers wrote in 2010, *"the genome is mostly dead, transcriptionally. The junk is still junk"* [Mye10].

While the mutation mechanism for gene formation makes an interesting story, there are a number of scientific difficulties with the scenario. Blind chance is the only known mechanism possible for such gene formation since a selective advantage was thought to require some manifestation that is genetically coded, such as making a particular protein. Since nothing prevents what would have become a mutated "correct" gene from mutating again to become useless (mutation is by chance), the probability for a useful mutated gene is that all required mutations take place in one organism before or during reproduction.

A typical gene contains over 1,000 base-pairs, but if we speculate that an operative gene could have only the codons required for encoding a 50-amino acid protein, the minimum length of the gene would be at least 150 base-pairs. The probability of forming such a gene can be estimated as $4^{-150} = 5 \times 10^{-91}$ (that's for one specific gene in one organism, which may decrease for larger gene size, which would be typical, or increase if additional functional genes or redundant coding are considered). Even this extremely small gene has a probability that is well below the operationally falsified cut-off [Abe09U]. Any new protein would also require additional promoter proteins to find the transcription-starting nucleotide, as well. Since "useless" base-pairs have no advantage and since transferring the information requires energy (at least 0.035 electron volts per bit) for each step during replication [Yoc05p25] (there's no free lunch!), any mutation that eliminated those useless base-pairs would have a selective advantage. It would make sense that those useless nucleotides would be removed from the genome long before they had a chance to form something with a selective advantage. Time is actually an enemy since entropy will tend to randomize genetic content unless directed energy prevents that from happening [Abe06], and there would be no advantage in directing energy to "useless" structures.

"Junk DNA" has long been classified as a misnomer by scientists doubting Darwinism [Den86], since *"junk DNA and directed evolution are in the end incompatible concepts"* [Den98]. The journal Science refused to print a 1994 letter warning about assuming that "junk DNA" was useless [Mim94]. Recent studies have shown that non-coding DNA plays a role in embryonic development in such areas as the reproductive tract [Kep96] and the central nervous system [Koh96]. It also is crucial in preventing heart disease and cancer [Fan10]. Non-coding DNA is also critical in regulating replication [Ste05] and for DNA repair [Mor02]. *"Biology's new glimpse at a universe of non-coding DNA — what used to be called 'junk' DNA — has been fascinating and befuddling"* [Hay10]. John Mattick, Director of the Centre for Molecular Biology and Biotechnology at the University of Queensland, writes: *"I think this will come to be a classic story of orthodoxy derailing objective analysis of the facts, in this case for a quarter of a century, the failure to recognize the full implications of this — particularly the possibility that the intervening noncoding sequences may be transmitting parallel information in the form of RNA molecules—may well go down as one of the biggest mistakes in the history of molecular biology... Indeed, what was damned as junk because it was not understood may, in fact, turn out to be the very basis of human complexity"* [Mat03]. See Appendix F for more details on "junk DNA," but realize that, since a protein is the result of the execution of a prescriptive algorithm, what had been assumed is that random memory changes can produce a new functional program.

Researchers are discovering that what had been dismissed as evolution's relics are actually vital to life. What used to be considered evidence of the neo-Darwinism gene-formation mechanism can no longer be used as such evidence. In this case, neo-Darwinism has been a proven science inhibitor as it postponed serious investigation of the non-coding DNA within the genome, which was *"one of the biggest mistakes in the history of molecular biology"* [Mat03]. This is reminiscent of the classification of 86 (later expanded to 180) human organs as "vestigial" that Robert Wiedersheim (1893) believed *"lost their original physiological significance,"* in that they were vestiges (footprints /traces) of evolution [Wie1893]. Functions have since been discovered for all 180 organs that were thought to be vestigial, including the wings of flightless birds, the appendix, and the ear muscles of humans [Ber90].

It should also be noted that just having a selective advantage wouldn't guarantee that such an advantage would continue to be inherited [Spe97p80]. Even with a 0.1% advantage, in a population of 1 million, the probability of survival of that mutant strain is only about 10^{-6}, whereas in a population of 100, the probability of survival increases to 0.011 (see Appendix A), which has led many to believe macroevolution only occurs in isolated groups of organisms in what is known as "punctuated equilibria" [Eld72]. In other words, just having a selective advantage does not mean the mutant would survive. Another problem is the cost of substitution that can occur in a given time as the favorable genes spread in the population by the lowered fertility (or death) of those without those genes [Hal57]. It would take time for even desirable mutations to become significant.

Functional information increasing as one traverses the tree upward from its root is another inexplicable feature of Darwinism. *"The pressure to code information in a fault-tolerant manner implies that codes should evolve that are robust to deleterious mutations*" [Ofr03]. There is no doubt that mutations can modify existing information within the genome to produce modified characteristics. Segment and point mutations can cause a loss of information, and that loss may actually increase survivability (but is usually the opposite).

For example, if a bacterium mutates at the point where an antibiotic would attach, it becomes resistant to that antibiotic. The "best" place to become infected with such bacteria is a hospital, because nearly all of the "normal" bacteria have been killed, leaving only the "deformed" bacteria to compete for survival. Sickle cell anemia is caused by a point mutation in the hemoglobin beta gene. While normally this defect is detrimental due to lowered oxygen-carrying capability, the plasmodium parasite that causes malaria has difficulty invading cells with that mutation, so those having this genetic defect are protected against malaria, which is a selective advantage in high malaria regions.

The nylon-digesting bacterium [Kin75] is often cited as an example of information increase via mutation, especially since nylon didn't exist before 1935. This may be the result of a "frame-shift" (with decoding off by one nucleotide) mutation that enables the bacterium to digest nylon at about 8% efficiency. Since it can no longer digest its normal diet of cellulose, a net functional information loss is evident in the genome.

"The human genome is a big document full of information, like a blueprint... In this study, we're looking at transposons that insert themselves in new places in various genomes and disrupt the blueprint... If you think of the human genome as a manual to build a complex machine like an aircraft, imagine what would happen if you copied the page that describes passenger seats and inserted it into the section that describes jet engines. Transposons act something like this: they copy themselves and insert the copies [not new information] *into other areas of the human genome, areas that contain instructions for the complex machine that is the human body. These areas and the instructions they*

contain may then become corrupted and hard to understand. This, in turn, can alter human traits or even cause human diseases" [Dev10].

"There is no evidence that genetic information can build up through a series of small steps of microevolution... Mutations reduce the information in the gene by making a protein less specific. They add no information, and they add no new molecular capability... None of them can serve as an example of a mutation that can lead to the large changes of macroevolution... The failure to observe even one mutation that adds information is more than just a failure to find support for the theory. It is evidence against the ... neo-Darwinian theory" [Spe97p159-160].

"Stunningly, information has been shown not to increase in the coding regions of DNA with evolution. Mutations do not produce increased information... the amount of coding in DNA actually decreases with evolution... No increase in Shannon or Prescriptive information occurs in duplication" [Abe09G]. A new species would require multiple new structures, each made up of proteins, each requiring an instantiated functional prescriptive algorithm.

Evolutionary geneticist H. Orr writes supporting punctuated equilibrium in some cases, *"We conclude – unexpectedly – that there is little evidence for the neo-Darwinian view: its theoretical foundations and the experimental evidence supporting it are weak"* [Orr92p726]. *"None of the papers published in JME (Journal of Molecular Evolution) over the entire course of its life (1971-) as a journal has ever proposed a detailed model by which a complex biochemical system might have been produced in a gradual, step-by-step Darwinian fashion"* [Beh96p176]. *"We must concede there are presently no detailed Darwinian accounts of the evolution of any biochemical or cellular system, only a variety of wishful speculations"* [Har01].

Biologist Lynn Margulis writes, *"We agree that very few potential offspring ever survive to reproduce and that populations do change through time, and that therefore natural selection is of critical importance to the evolutionary process. But this Darwinian claim to explain all of evolution is a popular half-truth whose lack of explicative power is compensated for only by the religious ferocity of its rhetoric. Although random mutations influenced the course of evolution, their influence was mainly by loss, alteration, and refinement. One mutation confers resistance to malaria but also makes happy blood cells into the deficient oxygen carriers of sickle cell anemics. Another converts a gorgeous newborn into a cystic fibrosis patient or a victim of early onset diabetes.*

One mutation causes a flighty red-eyed fruit fly to fail to take wing. Never, however, did that one mutation make a wing, a fruit, a woody stem, or a claw appear. Mutations, in summary, tend to induce sickness, death, or deficiencies. No evidence in the vast literature of heredity changes shows unambiguous evidence that random mutation itself, even with geographical isolation of populations, leads to speciation. Then how do new species come into being?" [Mar03]

Bacteriologist Alan Linton, looking for confirmed primary speciation states, *"None exists in the literature claiming that one species has been shown to evolve into another. Bacteria, the simplest form of independent life, are ideal for this kind of study, with generation times of twenty to thirty minutes, and populations achieved after eighteen hours. But throughout 150 years of the science of bacteriology, there is no evidence that one species of bacteria has changed into another"* [Lin01].

"Yes, small-scale evolution is a fact, but there is no reason to think it is unbounded. In fact, all our data suggests that small-scale evolution cannot produce the sort of large-scale change Darwinism requires" [Hun03]. At a Conference on Macroevolution, anthropologist Roger Lewin said, *"The central question of the Chicago conference was whether the mechanisms underlying microevolution can be extrapolated to explain the phenomena of macroevolution. At the risk of doing violence to the position of some people at the meeting, the answer can be given as a clear, No"* [Lew80].

The fossil record is believed by many to support neo-Darwinism, but there are many problems therein. The Cambrian explosion refers to the geologically sudden (within no more than 0.1% of earth's history) appearance of at least 19, or as many as 35 of the 40 total phyla (divisions of plant/animal kingdoms) of animals in the fossil record during the Cambrian period (lowest level with multi-cellular fossils).

During this event, each phylum exhibits a unique architecture, blueprint, or structural body plan (morphology) with no predecessors or intermediaries. Each physical change in structure would necessarily be the result of functional and prescriptive informational changes in the DNA. Examples of basic animal body plans include cnidarians (corals and jellyfish), mollusks (squids and shellfish), arthropods (insects, crustaceans, and trilobites), and the chordates (phylum of all vertebrates, including humans).

"Most of the animal phyla that are represented in the fossil record first appear, 'fully formed,' in the Cambrian... The fossil record is

therefore of no help with respect to the origin and early diversification of the various animal phyla" [Bar01]. *"If subphyla are included in the count of animal body plans, then at least thirty-two and possibly as many as forty-eight of fifty-six total body plans (57.1 to 85.7 percent) first appear on earth during the Cambrian explosion"* [Mey03p330]. *"All living phyla may have originated by the end of the explosion"* [Val99p327].

Microbiologist Carl Woese writes, *"Phylogenetic incongruities can be seen everywhere in the universal tree, from its root to the major branchings within and among the various taxa to the makeup of the primary groupings themselves"* [Woe98]. Though there are certainly minor changes in body structures, the sparsity of what are believed to be "transitional" forms, with none being unambiguous, is another problem if evolution is a small step at a time.

Paleontologist and evolutionary biologist Henry Gee notes, *"To take a line of fossils and claim that they represent a lineage is not a scientific hypothesis that can be tested, but an assertion that carries the same validity as a bedtime story—amusing, perhaps even instructive, but not scientific"* [Gee99]. Fossilized, animals fall clearly within one of a limited number of basic body plans with clear morphological differences [Hal96]. *"The animal body plans (as represented in the fossil record) do not grade imperceptibly one into another, either at a specific time in geological history or over the course of geological history. Instead, the body plans of the animals characterizing the separate phyla maintain their distinctive morphological and organizational features and thus their isolation from one another, over time"* [Mey03p333]. Dawkins writes concerning the invertebrate phyla fossils, *"It is as though they were just planted there, without any evolutionary history"* [Daw96Bp229].

"Evolution...must be gradual when it is being used to explain the coming into existence of complicated, apparently designed objects, like eyes... Without gradualness in these cases, we are back to miracle, which is simply a synonym for the total absence of explanation" [Daw95]. *"The problem of how eyes have developed has presented a major challenge to the Darwinian theory of evolution by natural selection. We can make many entirely useless experimental models when designing a new instrument, but this was impossible for Natural Selection, for each step must confer some advantage upon its owner, to be selected and transmitted through the generations. But what use is a half-made lens? What use is a lens giving an image, if there is no nervous system to*

interpret the information? How could a visual nervous system come about before there was an eye to give it information? In evolution there can be no master plan, no looking ahead to form structures which, though useless now, will come to have importance when other structures are sufficiently developed. And yet the human eye and brain have come about through slow painful trial and error" [Gre72].

There continues to be much speculation on the origin of the eye, ranging from multiple (as many as 60) independent times [Koz08] to a single evolved eye from which all eyes evolved [Geh05]. Even the simplest light sensitive spot involves a large number of specialized proteins and molecules in an extremely complicated integrated system. If any one of these proteins or molecules is missing in even the simplest eye system, there is no vision. A vision study of a *"combination of spectral measurements and genetic sequences revealed which DNA mutations – that is, amino acid substitutions – were responsible for the shift in peak wavelength. The results were surprising: identical substitutions didn't always produce the same shift... The results show how hard it is to identify productive mutations and to predict their effect. Showing conclusively how the survival of the fittest plays out on the molecular level would require reconstructing not only the protein, but also the whole animal and its long-lost habitat"* [Yok08]. These experiments demonstrated that even modifications that were "supposed" to be useful were very difficult to characterize, and in fact would require information no longer available. One of the characteristics of a scientific theory is the ability to make accurate predictions.

The trilobite eye is a good case study since the trilobites suddenly appeared in the Cambrian (lowest fossil-bearing) stratum with no record of ancestry. The trilobite eye is made of optically transparent calcium carbonate (calcite, the same material of its shell) with a precisely aligned optical axis that eliminates double images and two lenses affixed together to eliminate spherical aberrations [McC98, Gal00].

Paleontologist Niles Eldredge observed, *"These lenses — technically termed aspherical, aplanatic lenses — optimize both light collecting and image formation better than any lens ever conceived. We can be justifiably amazed that these trilobites, very early in the history of life on Earth, hit upon the best possible lens design that optical physics has ever been able to formulate"* [Eld76]. Notice these lenses weren't just as good as, but were better than anything modern optical physicists have been able to conceive! *"The design of the trilobite's eye lens could well qualify for a patent disclosure"* [Lev93p58]. See Appendix F for more details.

The trilobite lens is particularly intriguing since the only other animal to use inorganic focusing material is man. The lens may be classified as a prosthetic device since it was non-biological, which also means the lens itself (with no DNA) was not subject to Darwinian evolution. The manufacturing and controlling of the lenses were obviously biological processes, with an unknown number of DNA-prescribed proteins (each with a prescriptive manufacturing program) for collecting and processing the raw materials to manufacture the precision lenses and create the refracting interface between the two lenses.

The lenses do not decompose as any other animal's lenses would, so they are subject to rigorous scientific investigation and determination of optical properties based on the actual lenses, from which inferences can be made as to their use. For example, since they are multi-focal with spherical aberration correction, it can be inferred that trilobites had very good visual acuity. All of these attributes require that considerable new prescriptive information somehow is inserted into the genome in order to indirectly manufacture the lenses. Since no immediate precursors of trilobites have been found, Darwinists are without any evidence as to how an organism with an eye as complex as a trilobite could have arisen, especially in such a relatively short time in the lowest multi-cellular fossil-bearing stratum, near the very beginning of life.

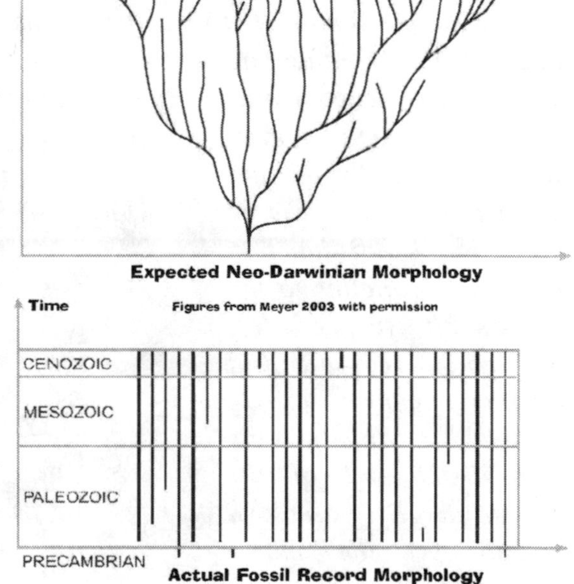

We see in the features of the Cambrian explosion, *"(1) a quantum or discontinuous increase in specified complexity or information; (2) a topdown pattern of innovation in which large-scale morphological disparity arises before small-scale diversity; (3) the persistence of structural (or "morphological") disparities between separate organizational systems; and (4) the discrete or simultaneous*

emergence of functionally integrated material parts within novel organizational body plans" [Mey03-p390]. It's interesting that the top-down structural changes observed in Cambrian strata mirror the top-down structuring required by software engineering.

Irreducible complexity is another problem that neo-Darwinian evolution has not addressed in any meaningful way. Behe defines irreducibly complex as, *"a single system composed of several well-matched, interacting parts that contribute to basic function, wherein the removal of any one of the parts causes the system to effectively cease functioning. An irreducibly complex system cannot be produced directly... by slight, successive modifications of a precursor system, because any precursor to an irreducibly complex system that is missing a part is by definition non-functional"* [Beh96p39]. A very sudden increase in prescriptive information would be required to build a complex machine that had no predecessor since each protein of the structure requires a functional prescriptive algorithm. Behe's examples include the flagellum, aspects of blood clotting, closed circular DNA, telomeres (condensed DNA material), photosynthesis, and transcription regulation.

The bacterial flagellum is one of the best known examples of an irreducibly complex system [Beh96]. It is an organelle in a single-cell organism. "*The bacterial flagellum of Salmonella typhimurium is the analogue of a man-made mechanical system. Its heart is a 15,000 revolutions per minute, reversible rotary motor powered by the proton-motive gradient across the cell's inner membrane. Each revolution consumes about 1000 protons. A drive shaft, held by a bushing in the outer membrane, transmits torque across the cell's envelope. Attached to the drive shaft, a universal joint enables the motor to drive the propeller, even when the drive shaft and propeller are not co-linear. A short junction joins the propeller to the drive shaft. The propeller, a long left-handed corkscrew, converts torque to thrust. A cap sits at the cell distal end of the filament. By electron microscopy, the motor*

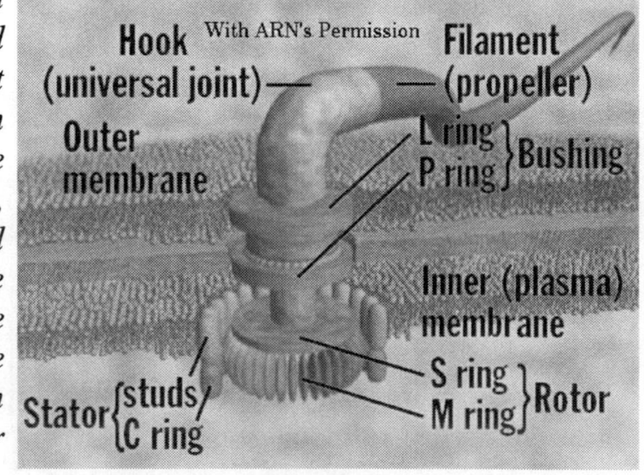

associated parts and the bushing are seen to be rings of subunits, whereas the drive shaft appears to be a helical assembly of subunits. About four dozen genes are needed to build the flagellum. Some are required for regulation of synthesis; some for export and assembly; some for the structure itself, and a few are of unknown function" [DeR95].

Over 40 distinct, carefully positioned proteins (each constructed from complex PI) are used to construct the flagellum, including the various parts (including O-rings, rotor, stator, and bushings) and connectors to non-flagellum parts. *"The bacterial flagellar motor is an example of finished bio-nanotechnology, and understanding how it works and assembles is one of the first steps towards making man-made machines on the same tiny scale. The smallest man-made rotary motors so far are thousands of times bigger...* [a flagellum] *spins at up to 100,000 rpm and achieves near-perfect efficiency"* [Ber06B].

"Flagellar proteins are synthesized within the cell body and transported through a long, narrow central channel in the flagellum to its distal (outer) end, where they self-assemble to construct complex nano-scale structures efficiently, with the help of the flagellar cap as the assembly promoter. The rotary motor, with a diameter of only 30 to 40 nm, drives the rotation of the flagellum at around 300 Hz, at a power level of 10^{-16} W with energy conversion efficiency close to 100 %. The structural designs and functional mechanisms to be revealed in the complex machinery of the bacterial flagellum could provide many novel technologies that would become a basis for future nanotechnology, from which we should be able to find many useful applications" [Nam02].

If any one of 40 proteins is missing, a complete loss of motor function results. ATP is not used as an energy source, but instead, bacteria use energy from the flow of ions across membranes. *"One can only marvel at the intricacy, in a simple bacterium, of the total motor and sensory system which has been the subject of this review and remark that our concept of evolution by selective advantage must surely be an oversimplification. What advantage could derive, for example, from a 'preflagellum' (meaning a subset of its components), and yet what is the probability of 'simultaneous' development of the organelle at a level where it becomes advantageous?"* [Mac78]

It has been noted that several proteins that are needed by the flagellum are also found in Type III Secretion Systems (TTSS) that bacteria use to inject poisons into victims, with the speculation that TTSS is a precursor of the flagellum. Although 10 genes are shared, 40 (30 in the motor) genes are unique to the flagellum. Estimation of evolutionary relationships among organisms on gene sequences *"suggest that flagellar motor proteins arose first and those of the pump came later... if anything, the pump evolved from the motor, not the motor from the pump"* [Min04]. The pump is used during flagellum construction since self-assembly is performed at the tip (distal end).

Since each protein requires a prescriptive algorithm implemented in the DNA, the simultaneous "arising" of at least 30 such computer programs is difficult to justify from information science. When we consider 10^{-164} as the probability of even a simple life-compatible protein [Mey09p212] (even forming one is operationally falsified [Abe09U]), the probability of forming all 30 becomes 10^{-4920}.

Dembski [Dem04] points out the daunting probabilistic hurdles Darwinians face when attempting coordination of the necessary successive evolutionary changes needed for biomechanical machines that are irreducibly complex, including availability and synchronization of parts, elimination of interfering cross-reactions and proper interfacing with other components, and ordering the assembly so that a functioning system results. John Bracht [Bra03], Michael Behe [Beh00], and William Dembski [Dem04] have adequately addressed issues raised by skeptics concerning irreducible complexity, especially that of the flagellum.

It should be noted that Yockey's statement, *"the protein sequences that compose living organisms are not random or 'irreducibly complex'"* [Yoc05p179], is technically true considering only Shannon complexity and not functional or prescriptive information. Irreducible complexity, as used by Behe, deals with forming new functional structures (an engineering problem, rather than a mathematical puzzle), along with the associated proteins that require considerable functional information.

New proteins typically *"require multiple new protein folds, which in turn require long stretches of new protein sequence.... the vast set of possible proteins that could conceivably be constructed by genetic mutations is far too large to have actually been sampled to any significant extent in the history of life. Yet how could the highly incomplete sampling that has occurred have been so successful? How could it have located the impressive array of protein functions required for life in all its forms, or the comparably impressive array of protein structures that perform those functions?... If we take 300 residues as a typical chain length for functional proteins, then the corresponding set of amino acid sequence possibilities is unimaginably large, having 20^{300}*

(= 10^{390}) members... [Mutational paths to create] *new folds will inevitably destabilize the original fold before producing the new one... the sampling problem— the impossibility of any evolutionary process sampling anything but a miniscule fraction of the possible protein sequences... greatly strengthened the case that the sampling problem is real and that it does present a serious challenge"* [Axe10].

Irreducible complexity may also be found in the system of DNA/RNA/enzymes. *"Nucleic acids are synthesized only with the help of proteins, and proteins are synthesized only if their corresponding nucleotide sequence is present. It is extremely improbable that proteins and nucleic acids, both of which are structurally complex, arose spontaneously in the same place at the same time. Yet it also seems impossible to have one without the other. And so, at first glance, one might have to conclude that life could never, in fact, have originated by chemical means"* [Org94].

Since the most popular books seem to involve sex, this book will do so also. Sexual reproduction may also be an example of irreducible complexity that Darwinism has failed to explain. The ubiquitous nature of sexual reproduction, whether in plants or animals, is based on gender differences and defies any naturalistic explanation.

How could a female member of a species evolve to produces eggs and be internally equipped to nourish a growing embryo, while at the same time a male member evolve to produce viable sperm, with each gamete (sperm or egg) evolving to contain half the number of chromosomes? Sex's *"disadvantages seem to outweigh its benefits. After all, a parent that reproduces sexually gives only one-half its genes to its offspring, whereas an organism that reproduces by dividing passes on all its genes. Sex also takes much longer and requires more energy than simple division. Why did a process so blatantly unprofitable to its earliest practitioners become so widespread?"* [Sch84]

"Despite decades of speculation, we do not know. The difficulty is that sexual reproduction creates complexity of the genome and the need for a separate mechanism for producing gametes. The metabolic cost of maintaining this system is huge, as is that of providing the organs specialized for sexual reproduction" [Mad98]. *"Sex is a puzzle that has not yet been solved; no one knows why it exists"* [Rid01].

Recombinant DNA [Coh73], the "creation" of new life forms in the laboratory, may raise questions whether such creations support neo-Darwinism. The artificial DNA is engineered through combination or insertion of one or more existing DNA strands to reprogram DNA to include sequences not normally occurring together. The modified or new traits are designed for a specific purpose, such as immunity or a new fuel [Ven08]. No new net information is generated, as existing information is used to form the modified information, with some of the original information lost in the new DNA.

As an analogy, this book has to this point not created any new information structures (words). If "juxtaword" is used in this paragraph, the reader may recognize the meaning. It's unclear how many words can be juxtaworded (using the principle than any noun can be verbbed), or the limits of juxtawording, or if multiple juxtawords may be juxtaworded. If people like the use of "juxtaword," it could spread to the population in general, becoming a permanent trait.

This is an example of a minor modification that could have been produced by juxtaposing the first five letters of the information unit "juxtapose" with the information unit "word." Note that the "pose" information is missing in "juxtaword." Information units are analogs to genes, with the letters being analogs to the genetic alphabet. If a totally new information unit "swervozt" is randomly generated from the alphabet, it's extremely unlikely that it would be replicated throughout the population, since it was produced without purpose, and in fact contains no functional information (although the Shannon complexity is higher than a normal word due to the unpredictability of the letters).

The "artificial genome" produced by the Venter Institute was unique in genetic engineering in that a genome that matched (except for a few additions, such as identifying watermarks) an existing bacterium was engineered from existing parts (all parts originated from life). This genome replaced the genome in another bacterium, leaving all the computing machinery, operating systems, and other components of that cell intact. To put this in perspective, copying the executable binary code of a program to a USB drive and then inserting the drive into another computer is not creating a system "from scratch." The new genome was engineered using computers *"starting from digitized genome sequence information,"*[Gib10] verifying the digital computing nature of life. The complexity and specificity of life's information is highlighted by *"obtaining an error-free genome that could be transplanted into a recipient cell to create a new cell controlled only by the synthetic genome was complicated and required many quality control steps. Our success was thwarted for many weeks by a single base pair deletion in the essential gene dnaA. One wrong base out of over one million in an essential gene rendered the genome inactive"*[Gib10]. One of the things this research supports is the idea that (at least for the two closely-related bacteria involved) life uses common operating systems, programming languages, and devices (otherwise the programs for one machine wouldn't execute on another).

While there is evidence that life increased in complexity with many new body structures over time, there is no scientific evidence that the increase in information required to make those changes could have occurred in a Darwinian manner, as is usually assumed. The microevolutionary changes that have been observed seem to have a very limited range to affect form or function, whereas the changes required for macroevolutionary changes have not been observed or are detrimental.

"Loci that are obviously variable within natural populations do not seem to lie at the basis of many major adaptive changes, while those loci that seemingly do constitute the foundation of many if not most major adaptive changes are not variable" [McD83]. *"The mechanism of heredity by gene action is insufficient to explain observations"* [Ede66].

The Darwinian model for evolution is increasingly being challenged by scientists. The observed *"multi-character changes are fundamentally different from the slowly accumulating small random variations postulated in Darwinian and neo-Darwinian theory... One of the traditional objections to Darwinian gradualism has been that it is too slow and indeterminate a process to account for natural adaptations, even allowing for long periods of random mutation and selection... natural genetic engineering... a cognitive component is absent from conventional evolutionary theory because 19th and 20th century evolutionists were not sufficiently knowledgeable about cellular response and control networks"* [Sha10]. Somehow a cell would do a combinatorial search of the possible outcomes of genetic changes in order to choose the path that would be most beneficial, which differs greatly from the purposelessness of Darwinian random mutations.

"Evolutionary-genomic studies show that natural selection is only one of the forces that shape genome evolution and is not quantitatively dominant, whereas non-adaptive processes are much more prominent than previously suspected. Major contributions of horizontal gene transfer and diverse selfish genetic elements to genome evolution undermine the Tree of Life concept. An adequate depiction of evolution requires the more complex concept of a network or 'forest' of life...

Evolutionary genomics effectively demolished the straightforward concept of the TOL by revealing the dynamic, reticulated character of evolution" [Koo09]. Another non-Darwinian mechanism involves Transposons, which *"are segments of DNA that can replicate themselves – meaning that each generation of a human family has more transposons in its genome than its ancestors"* [Dev10].

"Natural selection is the long-term result of molecular copying and would be the sole mechanism of evolution if copying were the sole basic mechanism of life. But there are two distinct molecular mechanisms at the basis of life, copying and coding, and both of them have long-term consequences. Copying leads in the long run to natural selection and coding to natural conventions [biosemiotics], *which means that evolution took place by two distinct mechanisms. Natural selection produces new objects by modifying previous ones, whereas natural conventions bring absolute novelties into existence"* [Bar08B]. New structures and new species would be the result of new arbitrary natural coding systems.

"Much of the vast neo-Darwinian literature is distressingly uncritical... Natural selection has shown insidious imperialistic tendencies" [Fod10S]. *"There is something wrong – quite possibly fatally wrong – with the theory of natural selection ... Neo-Darwinism is taken as axiomatic; it goes literally unquestioned"* [Fod10Wpviii-ix]. Biology Professor Scott Gilbert says, *"Natural selection occurs within species. But natural selection alone cannot explain how butterflies got their wings. How the turtle got its shell. Once you have variation within species, then natural selection can work. Do I think natural selection should be relegated to a less important role in the discussion of evolution? Yes I do"* [Maz10p220]. "*It is relatively well known how organisms adapt to their environment and, arguably, even how new species originate. However, whether this knowledge suffices to explain macroevolution, narrowly defined here to describe evolutionary processes that bring about fundamental novelties or changes in body plans has remained highly controversial"* [Hin06].

"Natural selection can't be the mechanism of evolution... Introducing mental states into the operation of natural selection would allow it to reconstruct the distinction between selection and selection-for... but the cost would be catastrophic. Mental processes require minds in which to happen" [Fod10Wp114&155].

NASA Astrobiology Institute Chief Bruce Runnegar says, "*Natural selection is not a mechanism, it's the process by which the results of evolution are sorted... all of the processes are much more complicated than people imagine. There are many more loops in the biochemistry of organisms. There are many cases where the RNA itself does the job and feeds back into the protein loop. So this whole system has become so much more complex. We understand the nature of life a lot more than we did 10 years ago"* [Maz10p188&190]. Presidential Medal of Science winner Lynn Margulis notes, *"as far as 'survival of the fittest' goes, ...* [natural selection is] *neither the source of heritable novelty nor the entire evolutionary process...* [making] *Darwinism 'dead,' since there's no adequate evidence in the literature that random mutations result in new species... Natural selection is the failure to reach the potential, the maximum number of offspring that, in principle, can be produced by members of the specific species in question. This has been shown zillions of times in zillions of organisms"* [Maz10p257&267].

It is extremely important to realize that neo-Darwinism is not a mechanism for evolutionary change! A mechanism would explain how new structures develop. Each structure is the result of multiple proteins, each requiring an instantiated prescriptive algorithm for its production. The finished structures of an organism that can contribute to survival selection are determined by the genetic changes. Darwinism offers no feasible mechanism for specifying the algorithms or other functional information at the genomic level (randomness is not a mechanism for generating functional information). Darwinism asserts that the fittest will survive, and that those that survive are the fittest, which is tautology that can be stated "those that survive, survive."

We know from artificial life programs that no positive evolutionary progress is made without appropriate fitness functions and targets. We also know that complexity of organisms has increased from the lowest strata to the present, with many unique structures that would require increased functional information in the genomes.

With the human genome having a functional information of over 10^8 bits (6×10^9 bits of total information), and the simplest life only 267,000 bits, somehow over 10^7 bits of functional information was injected into the genome. It is relatively easy to see how a genome can increase in length via insertion or replication, but such mutations would add almost no information. For example, the last three repeats in the string "Anythingnew?Anythingnew?Anythingnew?Anythingnew?," add no Shannon or functional information (except for a repeat count of 4).

There is no known physical mechanism for producing any increase in information, let alone functional information of the magnitude necessary to evolve from the root to the top of life's tree. The ratios of probabilities of forming genomes of a human to that of the simplest organism, using the functional information formula, would be $2^{-10,000,000} = 10^{-3,000,000}$, which is clearly infeasible.

Chapter 9 will address the concerns and unanswered questions in more detail, but realize that neo-Darwinism is insufficient to explain the observations. Perhaps someday a scenario will suffice, but with our current knowledge, the scenarios proposed are severely lacking in explanatory capability. It will be interesting to see which of the many proposed scenarios will replace neo-Darwinism as naturalism's "truth" for the origin of species. Any acceptable scenario needs to be compatible with information science since it is information science, and not physical science, that determines life.

9 Unresolved Difficulties of Life's Information Requirements

"The failure of purely physical theories to describe or explain information reflects Shannon's concept of entropy and his measure of 'news.' Information is defined by its independence from physical determination: if it is determined, it is predictable and thus by definition not information. Yet Darwinian science seemed to be reducing all nature to material causes" [Gil06]. *"A certain wish fulfillment emerges from our naturalistic metaphysical presuppositions that uncontrolled physicodynamic phenomena will spontaneously self-organize into extra-ordinary degrees of formal ingenuity. Empirical support, logic, and prediction fulfillment evidence is sorely lacking for this blind, unfalsifiable belief"* [Abe10].

Michael Polanyi turned to philosophy at the height of his scientific career when he saw how ideologies were being employed to hinder free scientific expression and inquiry. Polanyi argued that life is not reducible to physical and chemical principles, but rather that, *"the information content of a biological whole exceeds that of the sum of its parts"* [PolWeb]. *"A whole exists when it acts like a whole, when it produces combined effects that the parts cannot produce alone"* [Cor02].

The argument for abiogenesis (life from non-life) *"simply says it happened. As such, it is nothing more than blind belief. Science must provide rational theoretical mechanism, empirical support, prediction fulfillment, or some combination of these three. If none of these three are available, science should reconsider that molecular evolution of genetic cybernetics is a proven fact and press forward with new research approaches which are not obvious at this time"* [Tre04]

"Biological functionality is turning out to be much more highly specified and precise than we had originally envisioned... biology is really a science of engineering, where the constraints for bio-functionality are extreme – to the point that nearly every molecular interaction is remarkably precise and tightly controlled. Molecular biology is much like a jigsaw puzzle where each piece must be specifically shaped to fit with the other pieces around it" [Bra03].

It is clear that every biologically functional part of an organism results from the execution of one or more prescriptive algorithms stored in the DNA memory of a cell. Thus far, no feasible physical explanation has been proposed as the cause of that prescriptive information or the other complex functional information within life, even though purely physical naturalism continues to be asserted as fact by many scientists.

Dean Kenyon published "Biochemical Predestination" [Ken69], an origin of life scenario proposing that natural chemical evolution could explain life. Kenyon has recanted this view, acknowledging that chemical forces cannot explain the information in life. Such predestination is still rampant in people's thinking, however. Billions have been spent for the search for life elsewhere, since a prevailing belief is that if life happened on Earth, it "must have" happened elsewhere, also. When evidence of water was discovered on Mars, the search for life on that planet intensified since "if there's water, there must be life."

Perhaps the time has come to reevaluate what we know, what we don't know, what's unknowable, and what is worth spending our limited resources to attempt to know. The last item is somewhat philosophical, but the others can be addressed as science. There is a need to differentiate what may be productive from speculations that can be operationally falsified [Abe09U]. So far, no origin of life (OOL) scenario has passed this feasibly test, but someday a scenario may, so that it wouldn't be ruled out by science. That wouldn't necessarily make it true, it just couldn't be falsified. For example, if the scenario had a 10^{-70} probability, it wouldn't be falsified even though it would be 10^{70} times more likely to be false than true. The result is that OOL is scientifically unknowable, so that any belief as to its source is just that, a belief.

Many scientists go to extraordinary lengths to protect Darwinism from dissent. The response to questioning Darwinism, for example, has often been loss of funding or even jobs [Ber08]. After considering the facts of this book, you can evaluate the accuracy of statements like *"Darwin's theory is now supported by all the available relevant evidence, and its truth is not doubted by any serious modern biologist"* [Daw82]. You can evaluate the claim that anyone doubting Darwinism *"is ignorant, stupid, or insane"* [Daw89]. This chapter introduces unscientific assumptions and beliefs that are often presented as science.

For example, although everyone acknowledges the reality of chance and necessity (law), many physicalists refuse to acknowledge the reality of formalism except for formal laws of physics. Those laws formally characterize expected results using mathematical equations, but there is no natural explanation for the source of those formal laws (they just "are"). Similarly, no feasible physical source has been proposed for the information formalities (verifiable scientific realities that just "are") described in this book. To present a scenario as science when it violates information science is not science, but faith. Science needs explanations.

Scientific dogmatism is counterproductive, even if widespread. Jonathan Wells [Wel00] writes of 10 evolutionary "Icons" that even evolutionists admit are misrepresentations, yet continue to be used as evidence of evolution. These icons include the Miller-Urey experiment, Darwin's tree of life, Haeckel's embryos, archaeopteryx, peppered moths, Darwin's finches, and ape-to-man evolution. Unfortunately, *"the commercial media is both ignorant of and blocks coverage of stories about non-centrality of the gene because its science advertising dollars come from the gene-centered Darwin industry... Thus, the public is unaware that its dollars are being squandered on funding of mediocre, middlebrow science or that its children are being intellectually starved as a result of outdated texts and unenlightened teachers"* [Maz10pix].

There is a decided (badly-needed) shift by many scientists away from neo-Darwinism (also called Modern Synthesis or MS), especially its gene centrism. *"The gene as the sole agent of variation and unit of inheritance, and the dogmatic insistence on this stance by the popularizers of the Synthesis, quelled all calls for more comprehensive attitudes. Although gene centrism has been a major point of contention, including strong criticism from philosophy of science, this aspect could not be changed from within the paradigm of the MS, which rested on it both explicitly and implicitly. But gene centrism necessarily disappears in an extended account that provides for multicausal evolutionary factors acting on organismal systems' properties, including the non-programmed components of environment,...* [with] *less overall weight to genetic variation as a generative force"* [Pig10]. *"Because biological systems are complex, a unified theoretical framework that coordinates, integrates, and even partially embeds a plurality of theories about systems is required to capture and manage this complexity... any single theoretical abstraction would lead to dangerous reifications* [treating abstract formalisms as if they were material]... *Ignoring key and legitimate abstractions can lead to limited understanding, short-sightedness, and the stalling of theoretical and empirical research*" [Win08].

While it is healthy to consider all alternatives, it should be pointed out that all materialistic scenarios proposed thus far fail to account for all information science principles needed. For example, horizontal gene transfer (HGT, gene transfer from one organism to another) doesn't violate those principles, assuming appropriate protocol and mechanism can be established. HGT maintains a key software engineering principle of "code reuse," since it is a standard practice for software engineers to lift working code from one program and insert it into a new program that has the same requirements. HGT, while not ruled out by information science, does not generate new information, however. It is not, therefore, a mechanism for generating new novelties. It may explain genetic similarities (homology) in different organisms, or even structural similarities (with an appropriate protocol and transfer mechanism).

"Science" has its root in the Latin word "Scientia," meaning knowledge or truth. *"Science... accumulated and accepted knowledge that has been systematized and formulated with reference to the discovery of general truths or the operation of general laws: knowledge classified and made available in work, life, or the search for truth: comprehensive, profound, or philosophical knowledge; especially knowledge obtained and tested through the use of the scientific method"* [Web93].

Nobel laureate Linus Pauling said, *"Science is the search for truth, the effort to understand the world; it involves the rejection of bias, of dogma, of revelation, but not the rejection of morality... One way in which scientists work is by observing the world, making note of phenomena, and analyzing them"* [PauWeb]. Appendix H highlights unscientific assumptions and beliefs that are often overlooked or not acknowledged as being valid (and that indirectly influence acceptance of this book's content). Since those beliefs inhibit scientific inquiry, it is important that they be exposed. Alternative beliefs won't be endorsed since they would also fall outside testability for scientific verification.

We have seen that all known science requires that components, including DNA, RNA, enzymes, and many other proteins must all be present and functional for life. We have seen that life simply cannot be determined by the laws of chemistry and physics alone, but requires formal control. We have seen the vast information content of even the simplest life, and that it is impossible for life to have arisen "proteins first." We know that formal thought can direct the production of extremely complex artifacts, such as this book, computers, or computer programs, with planning and future goals in mind. How could nature form artifacts whose complexity dwarfs anything man-made?

We have examined both the functional (especially prescriptive) information and the Shannon complexity of life, with Shannon information placing limits on information transfer, including the channel capacity limit that requires an initial alphabet of life to be at least as complex as the current DNA codon alphabet. We have seen that the complex functional information of life requires that it be intentionally prescribed.

We have seen irreducibly complex biological artifacts that require all parts, each manufactured via prescriptive algorithmic information, to simultaneously appear in an organism. We have looked at the Cambrian explosion to see a sudden influx of functional complexity for the formation of features in most (if not all) phyla, with no predecessor fossils. We have examined the recent findings of "junk DNA," which Darwin opponents would have encouraged examining over 20 years sooner, rather than dismissing it as evidence for Darwinian evolution.

For abiogenesis, we could consider 10^{-164} as the probability of forming a simple life-compatible protein by physicality [Mey09p212], or the simplest form of living organism known as $10^{-340,000,000}$ [Mor79]. These figures make these scenarios operationally falsified [Abe09U]. There really is no need for them, however, since the probability of a purely physical source of information contained in life is 0 (impossible, according to information science) based on alphabet requirements for information transfer [Sha48, Yoc05p182] and the complex coding and prescriptive cybernetic information processing systems in life (see Appendix G).

It is infeasible to put a reliable estimate on the probability of the production of the tree of life by Darwinian causes, considering the Cambrian explosion, irreducible complexity, etc. Also remember that no new net functional information, especially prescriptive information, results by mutations. Since the functional information of human DNA is over 10^7 bits more than the simplest organism, somehow over 10^7 bits would have to be injected into the genome, but there's no known information science mechanism for any net functional information increase. Using functional information, formation of the simplest organism can be shown to be $10^{3,000,000}$ times more probable than a human.

Known facts of life include its extreme cybernetic complexity, with millions of interacting co-dependent structures and components. Life is cybernetic in that it generates and controls its components using its components. Life's control and communication is digitally-based, and can be analyzed as a multi-computer system. Some of the specific problems that require explanation before propagating naturalistic speculations as science include the following.

How did nature write the prescriptive programs needed to organize life-sustaining metabolism? Programs are shown by computer science to require a formal solution prior to implementation. How did inanimate nature formally solve these complex problems and write the programs? How did nature develop the operating systems and programming languages to implement the algorithms? How did nature develop Turing machines capable of computational halting? How did nature develop the arbitrary protocols for communication and coordination among the thousands (or millions) of computers in each cell?

How did nature develop multiple semiotic coding systems, including the bijective codon-based coding system (for symbolic translation) that involves transcribing, communicating, and translating the symbolic triplet nucleotide block-codes into amino acids of the proteins? How did nature develop alternative generation of such messages using techniques such as overlapping genes, messages within messages, multi-level encryption, and consolidation of dispersed messages? A protein may obtain its consolidated prescriptive construction instructions from multiple genes and/or from the "junk" DNA, sometimes with over a million nucleotides separating the instructions to be combined.

How did nature defy computer science principles by avoiding software engineering's top-down approach required for complex programming systems? How did nature produce complex functional programs without planning by randomly modifying existing algorithms? How did multiple such programs become simultaneously modified to result in the production of irreducibly complex structures?

Speculation is important when generating or imagining new scientific theories. Scientists should strive to keep such speculation within the scientific community, however. Since the public tends to view what a scientist expresses as a view or belief as being "truth," it is important not to propagate unsubstantiated speculation as something worthy of consideration by non-scientists. It should be noted that all science is tentative, so new findings may require modification of what is considered "true."

The questions raised in this book require scientific answers before promoting as "science" any scenario for the origin of life or the origin of species. Perhaps different avenues of thinking are required in order to find scientific "truth" in these areas. Doggedly insisting that a scenario is true despite the evidence is unscientific. Maybe it's time to leave the "flat-earth" mentality that views things only from a particular limited perspective, and really examine what science is telling us.

References

Abel (David) & Jack Trevors, "Three Subsets of Sequence Complexity and Their Relevance to Biopolymeric Information," Theoretical Biology and Medical Modelling, 8/11/05, 2:29.

Abel (David) & Jack Trevors, "Self-Organization vs Self-Ordering Events in Life-Origin Models," Physics of Life Reviews (3), 2006, p211-228.

Abel (David), "Complexity, Self-organization, and Emergence at the Edge of Chaos in Life-Origin Models," Journal of the Washington Academy of Sciences 93 (4), 2007, p1-20.

Abel (David), "The 'Cybernetic Cut': Progressing from Description to Prescription in Systems Theory," The Open Cybernetics and Systemics Journal (2), 2008, p252-262.

Abel (David), "The GS (genetic selection) Principle," Frontiers in Bioscience (14), 1/1/09G, p2959-2969.

Abel (David), "The Biosemiosis of Prescriptive Information," Semiotica, 1/4/09P, p1-19.

Abel (David), "The Capabilities of Chaos and Complexity," Int. J. Mol. Sci. (10), 2009C, p247-291.

Abel (David), "The Universal Plausibility Metric (UPM) & Principle (UPP)," Theoretical Biology and Medical Modelling, 12/3/09U, 6:27.

Abel (David), "Constraints vs Controls," The Open Cybernetics & Systemics Journal (4), 2010, p14-27.

Adami (C.), "Ab Initio Modeling of Ecosystems with Artificial Life," Natural Resource Modeling: 15, 2002, p133-146.

Adleman (Leonard), "Molecular Computation Of Solutions To Combinatorial Problems," Science: 266, 11/11/94, p1021–1024.

Alberts (Bruce), Molecular Biology of the Cell, 1994, p533.

Ans-Web, http://www.answers.com/topic/overlapping-genes

Appeals Court, 7th Circuit, Kaufman, James v. McCaughtry & Gary, 8/20/05.

Appeals Court, Comer v. TEA, 5th Circuit, 7/2/10.

Astrobiology Magazine, "Inevitability Beyond Billions," 7/03.

Avida Manual, www.krl.caltech.edu/~charles/avida/manual/intro.html

Avida Website, www.krl.caltech.edu/avida/home/software.html

Axe (Doug), "Estimating the Prevalence of Protein Sequences Adopting Functional Enzyme Folds," J Mol Biol. 8/27/04, p1295-1315.

Axe (Doug), "The Case Against a Darwinian Origin of Protein Folds," BIO-Complexity, 2010(1), p1-12.
Ayala (Francisco), "Darwin's Revolution" in Creative Evolution?!, 1994, p3-5.
Babaoglu (O.), M. Jelasity, G. Canright, T. Urnes, A. Deutsch, N. Ganguly, G. Di Caro, F. Ducatelle, L. Gambardella, R. Montemanni, & A. Montresor, "Design Patterns from Biology for Distributed Computing," ACM Trans on TAAS, 5/9/06, p26-66.
Babbage (H. P.), "Babbage's Analytical Engine," Notices of the Royal Astronomical Society (70), 4/8/1910, p517–526.
Babraham Institute, "New Insight Into Reprogramming of Cell Fate," ScienceDaily, 2/1/10, sciencedaily.com/releases/2010/01/100127111105.htm
Bada (Jeffrey), "How Life Began on Earth: a Status Report," Earth and Planetary Science Letters: 226, 9/30/04, p1-15.
Bandyopadhyay (A.), R. Pati, S. Sahu, F. Peper & D. Fujita, "Massively Parallel Computing on an Organic Molecular Layer," Nature Physics 6, 4/25/10, p369-375.
Barash (Y.), J. Calarco, W. Gao, Q. Pan, X. Wang, O. Shai, B. Blencowe, & B. Frey, "Deciphering the Splicing Code," Nature, 5/6/10, p53-9.
Barbieri (Marcello), "Biosemiotics: a New Understanding of Life," Naturwissenschaften (95), 2/19/08B, p577-599.
Barbieri (Marcello), "Life is Semiosis – The Biosemiotic View of Nature," Cosmos and History: The Journal of Natural and Social Philosophy (4), 2008S, p29-51.
Barnes (R. K.), P. Calow & P. W. Olive, The Invertebrates: A New Synthesis, 2001, p9–10.
Barricelli (Nils Aall), "Esempi numerici di processi di evoluzione," Methodos, 1954, p45-68.
Barricelli (Nils Aall), Symbiogenetic Evolution Processes Realized by Artificial Methods, Methodos, 1957, p143–182.
Barry (Patrick), "Life from Scratch," Science News Online: 173 (2), 1/12/08, p27.
Battail (Gerard), "Genetics as a Communication Process Involving Error-Correcting Codes," in Biosemiotics: Information, Codes and Signs in Living Systems, 2007, p105.
Begley (Sharon), "Adventures in Good and Evil," Newsweek, 5/4/09, p46-48.
Behe (Michael), Darwin's Black Box: the Biochemical Challenge to

Evolution, 1996.
Behe (Michael), "Irreducible Complexity and the Evolutionary Literature: Response to Critics," 7/31/00, www.arn.org/docs/behe/mb_evolutionaryliterature.htm
Benenson (Yaakov), Binyamin Gil, Uri Ben-Dor, Rivka Adar, & Ehud Shapiro (2004-04-28), "An Autonomous Molecular Computer for Logical Control of Gene Expression," Nature: 429, 4/28/04, p423–429.
Benner (S.), H. Kim, M. Kim, & A. Ricardo, "Planetary Organic Chemistry and the Origins of Biomolecules," Cold Spring Harb Perspect Biol, 6/1/10.
Bennett (C. H.), "Logical Reversibility of Computation," IBM Jour of Research & Development: 17, 1973, p525-532.
Bergman (Jerry), http://www.khouse.org/articles/1997/143/, 1997.
Bergman (Jerry), Vestigial Organs Are Fully Functional, 1990.
Bergman (Jerry), Slaughter of the Dissidents: The Shocking Truth about Killing the Careers of Darwin Doubters, 2008.
Bernstein (Max), Jason Dworkin, Scott Sandford, George Cooper, & Louis Allamandola, "Racemic Amino Acids from the Ultraviolet Photolysis of Interstellar Ice Analogues," Nature: 416, 3/28/02, p401-403.
Bernstein (Max), "Prebiotic Materials From on and off the Early Earth," Phil. Trans. R. Soc. B:361, 2006P, p1689–1702.
Berry (Richard), in "Bacterial Motors Could Inspire Nanotechnology," 2/20/06B, www.physorg.com/news11029.html
Biomass, Wikipedia, http://en.wikipedia.org/wiki/Biomass_(ecology)
Boneh (Dan), Christopher Dunworth, Richard Lipton, and Jiri Sgall, "On the Computational Power of DNA," DAMATH 11, 1996, www.dna.caltech.edu/courses/cs191/paperscs191/bonehetal.pdf
Borek (Ernest), The Sculpture of Life, Columbia Univ Press, 1973, p5.
Borel (Emil), Probability and Certainty, 1950.
Borman (Stu), "Protein Factory Reveals Its Secrets," Chem & Eng News: 85(8), 2/19/2007, p13-16.
Bracht (John), "The Bacterial Flagellum: A Response to Ursula Goodenough," 2003, www.iscid.org /papers/Bracht_GoodenoughResponse_021203.pdf
Bruni (L.), "Cellular Semiotics and Signal Transduction," in Introduction to Biosemiotics, 2007, p365-407.
Buchanan (Mark), "Horizontal and Vertical: the Evolution of Evolution," New Scientist, 1/26/10, #2744.

Calderone (Melissa), "Do You Use More Energy When You're Thinking Really Hard?," www.popsci.com/scitech/article/2006-07/mental-workout

Cannarozzi (G.), N. Schraudolph, M. Faty, P. von Rohr, M. Friberg, A. Roth, P. Gonnet, G. Gonnet, & Y. Barral, "Role for Codon Order in Translation Dynamics," Cell 141, 4/16/10, p355-367.

Chaitin (Gregory), "Toward a Mathematical Definition of Life," in The Maximum Entropy Formalism, 1979.

Chaitin (Gregory), "Speculations on Biology, Information and Complexity," EATCS Bulletin: 91, 2/07, p231-237.

Cheng (Leslie) & Peter Unrau, "Closing the Circle: Replicating RNA with RNA," Cold Spring Harb Perspect Biol doi: 10.1101/cshperspect.a002204, 6/16/10.

Clarkson (Euan) & Riccardo Levi-Setti, "Trilobite Eyes and the Optics of Des Cartes and Huygens," Nature: 254, 4/24/75, p663-667.

Cohen (S.), A. Chang, H. Boyer, & R. Helling, "Construction of Biologically Functional Bacterial Plasmids In Vitro," PNAS USA: 70 (11), 11/73, p3240-3244.

Cooper (G.), N. Kimmich, W. Belisle, J. Sarinana, K. Brabham, & L. Garrel, "Carbonaceous Meteorites as a Source of Sugar-related Organic Compounds for the Early Earth," NASA Tech Report 20040088530, 2001.

Corning (Peter) & Stephen Kline, "Thermodynamics, Information and Life Revisited, Part I: to Be or Entropy," Systems Research, 4/7/00, p273-295.

Corning (Peter), "The Re-emergence of 'Emergence': A Venerable Concept in Search of a Theory," Complexity 7(6), 2002, p18-30.

Corning (Peter), Holistic Darwinism, 2005, p330.

Crews (Frederick),"Saving Us from Darwin," NY Review of Books: 48 (15), 10/4/01.

Crick (Francis), "The Origin of the Genetic Code," J Mol Biol: 38, 1968, p367–379.

Darnell (J.), H. Lodish, & D. Baltimore, Molecular Cell Biology, 1986.

Darwin (Charles), Origin of Species, Paperback (of 1859), 1998, p154.

Davenport (John), "Possible Progenitor of DNA Re-Created," Science Now, 11/16/00, p1.

Davis (Jimmy) & Harry Poe, Designer Universe, 2002.

Dawkins (Richard), The Selfish Gene, 1976.

Dawkins (Richard),"The Necessity of Darwinism," New Scientist: 94, 4/15/82, p130.

Dawkins (Richard), Scientific American, 6/88.
Dawkins (Richard), "Book Review," The New York Times, 4/9/89, section 7, p3.
Dawkins (Richard), River Out of Eden,1995, p83.
Dawkins (Richard), The Blind Watchmaker, 1996B.
Dawkins (Richard), Climbing Mount Improbable, 1996C.
Dawkins (Richard), A Devil's Chaplain: Reflections on Hope, Lies, Science, and Love, 2001, p79.
Dawkins (Richard), A Devil's Chaplain: Reflections on Hope, Lies, Science, and Love (paperback), 2004, p99.
Dawkins (Richard), "The Information Challenge," 1998 & 2008I, www.skeptics.com.au/articles/dawkins.htm
Dawkins (Richard), "Lying for Jesus?,"3/23/08L, http: //richarddawkins.net /article,2394,Lying-for-Jesus,Richard-Dawkins.
De Duve (C.),"The Beginning of Life on Earth," American Scientist, 1995.
Dembski (William), "Irreducible Complexity Revisited," 2004, www. designinference.com/documents/2004.01.Irred_Compl_Revisited.pdf
Denton (Michael), Evolution: A Theory in Crisis, 1986.
Denton (Michael), Nature's Destiny: How the Laws of Biology Reveal Purpose in the Universe, 1998.
DeRosier (D.), "Spinning Tails," Curr Opin Struct Biol., 4/5/95, p187-93.
Devine (Scott), quoted in " 'Jumping Genes' Are Known to Cause Disease," Science Daily, 6/25/10.
Diaconis (Persi), Susan Holmes, & Richard Montgomery, "Dynamical Bias in the Coin Toss," SIAM Review: 49 (2), 4/07, p211-235.
Dissent-web, "A Scientific Dissent From Darwinism," www.dissentfromdarwin.org/.
D'Onofrio (David) & Gary An, "A Comparative Approach for the Investigation of Biological Information Processing: an Examination of the Structure and Function of Computer Hard Drives and DNA," Theoretical Biology and Medical Modelling, 2010, 7:3.
Dose (Klaus), "The Origin of Life: More Questions Than Answers," Interdisciplinary Science Reviews: 13 (4), 1988, p348.
Durston (Kirk), David Chiu, David Abel, & Jack Trevors, "Measuring the Functional Sequence Complexities of Proteins," Theoretical Biology and Medical Modelling, 12/6/07, 14 pgs.
Easterbrook (Gregg), "Where did life come from?," Wired Magazine, 2/07, p108.

Eden (Murray), "Inadequacies of Neo-Darwinian Evolution as a Scientific Theory," Mathematical Challenges to the Neo-Darwinian Interpretation of Evolution, Wistar Institute, 1966.
Edinburgh University, "Insight into Cells Could Lead to New Approach to Medicines," Science Daily, 6/22/10.
Eiben (A.) & J. Smith, Introduction to Evolutionary Computing, 2003.
Ehrlich (P.) & L. Birch, "Evolutionary History and Population Biology," Nature, 4/22/67, p352.
Eldredge (Niles) & Stephen Jay Gould, "Punctuated Equilibria: an Alternative to Phyletic Gradualism," in T.J.M. Schopf, ed., Models in Paleobiology,1972 p82-115.
Eldredge (Niles), "A Trilobite Panorama in Eastern North America," Fossils Magazine: 1, 1976, p58-67.
Emmeche (Claus), "Defining Life, Explaining Emergence," 1997, www.nbi.dk/~emmeche/cePubl/97e.defLife.v3f.html
Entrez Database, www.ncbi.nlm.nih.gov/sites/entrez?db=protein
ERPANET/CODATA Workshop, "The Selection, Appraisal and Retention of Digital Scientific Data," Biblioteca Nacional, Lisbon, 12/15-17/03.
Evolution-site, http://evolution.berkeley.edu/evosite/evo101/IIntro.shtml.
Ewert (W.), G. Montañez, W. Dembski, R. Marks II, " Efficient Per Query Information Extraction from a Hamming Oracle," Proceedings of the the 42nd Meeting of the Southeastern Symposium on System Theory, IEEE, University of Texas at Tyler, 3/7-9/10, p290-297.
Fang (Janet), "Junk DNA Holds Clues to Heart Disease," Nature, 2/21/10, 10.1038/news.2010.82.
Faria (M.), "RNA as Code Makers: A Biosemiotic view of RNAi and Cell Immunity," in Introduction to Biosemiotics, 2007, p 347-364.
Flynn (M.), "Some Computer Organizations and Their Effectiveness," IEEE Trans. Comput: C-21, 1972, p948 .
FM-tribolites, www.fossilmuseum.net/Evolution/TrilobiteArmsRace.htm
Fodor (J.) & M. Piattelli-Palmarini, "Survival of the Fittest Theory: Darwinism's Limits," New Scientist, 2/3/10S.
Fodor (J.) & M. Piattelli-Palmarini, What Darwin Got Wrong, 2010W.
Ford (Brian), "On Intelligence in Cells: The Case for Whole Cell Biology," Interdisciplinary Science Reviews 34 (4), 12/09, p350-365.
Freeland (Stephen), Robin Knight, Laura Landweber, & Laurence Hurst, "Early Fixation of an Optimal Genetic Code," Molecular

Biology and Evolution: 17, 2000, p511-518.
Frey (J.), quoted in "Researchers Crack 'Splicing Code,' Solve a Mystery Underlying Biological Complexity," 5/5/10, www.physorg.com/news192282850.html
Gal (J.), G.Horvath, E. Clarkson, & O. Haiman, "Image formation by Bifocal Lenses in a Trilobite Eye?," Vision Research: 40, 2000, p843–853.
Gange (Robert), Origins and Destiny, 1986, p77.
Gates (Bill), The Road Ahcad, [1995], Revised, 1996, p.228.
Geddes (Linda), "Life's Code Rewritten in Four-letter Words," New Scientist, 2/17/10, #2748.
Gee (Henry), In Search of Deep Time, 1999, p113-117.
Gehring (W. J.), "New Perspectives on Eye Development and the Evolution of Eyes and Photoreceptors," Journal of Heredity: 96 (3), 2005, p171-184.
Geoclassics, "Trilobites," www.geoclassics.com/trilobites.htm.
Gibbs (W.), "The Unseen Genome: Gems Among the Junk," Scientific American, 11/03, p46-53.
Gibson (D.), J. Glass, C. Lartigue, V. Noskov, R. Chuang, M. Algire, G. Benders, M. Montague, L. Ma, M. Moodie, C. Merryman, S. Vashee, R. Krishnakumar, N. Assad-Garcia, C. Andrews-Pfannkoch, E. Denisova, L. Young, Z. Qi, T. Segall-Shapiro, C. Calvey, P. Parmar, C. Hutchison III, H. Smith, C. Venter, "Creation of a Bacterial Cell Controlled by a Chemically Synthesized Genome," Science Express, 5/20/10, p1-12.
Gilder (George), "Evolution and Me," National Review, 7/17/06.
Gimona (M.), "Protein Linguistics and the Modular Code of the Cytoskeleton," in The Codes of Lifc: The Rules of Macroevolution, 2008, p189-206.
Gingeras (Thomas), www.desdeelexilio.com/2010/06/28/epigenetica-entrevista-a-thomas-gingeras/
Gitt (Werner), In the Beginning was Information, 1997.
Glass (J.), N. Assad-Garcia, N. Alperovich, S. Yooseph, M. Lewis, M. Maruf, C. Hutchison III, H. Smith, & J. Venter, "Essential Genes of a Minimal Bacterium," PNAS: 103, 2006, p425-430.
GMIS (US Energy Department Genome Management Information System), http://genomics.energy.gov.
Gon (S. M.), "The Trilobite Eye," 10/1/07, www.trilobites.info/eyes.htm
Grassé (Pierre-P), Evolution of Living Organisms, 1977.

Gregory (R. L.), Eye and Brain: The Psychology of Seeing, second edition, 1972, p25.
Grünwald (David) and Robert Singer, "In Vivo Imaging of Labelled Endogenous (-actin mRNA During Nucleocytoplasmic Transport," Nature, 2010, DOI: 10.1038/nature09438.
Haeckel (Ernst), The Evolution of Man, Translation (original 1911) 2004, p49.
Haldane J.B.S.), "The Cost of Natural Selection," J. Genet.: 55, 1957, p511-524.
Hall (Brian), "Baupläne, Phylotypic Stages, and Constraint: Why There Are So Few Types of Animal," Evolutionary Biology: 29, 1996,
Hannon, (Gregory), quoted in "Messenger RNAs Are Regulated in Far More Ways than Previously Appreciated," Science Daily, 6/25/10.
Harold (F.), The Way of the Cell: Molecules, Organisms and the Order of Life, 2001, p205.
Hayden (Erika), "Human Genome at Ten: Life is Complicated," Nature 464, 3/31/10, p664-667.
Hazen (Robert), Patrick Griffin, James Carothers, & Jack Szostak, "Functional Information and the Emergence of Biocomplexity," PNAS: 104-1, 5/15/07, p8574-8581.
Hazen (Robert) & Dimitri Sverjensky, "Mineral Surfaces, Geochemical Complexities, and the Origins of Life," Cold Spring Harb Perspect Biol, 4/14/10, 2:a002162.
Hintz (M.), C. Bartholmes, P. Nutt, J. Ziermann, S. Hameister, B. Neuffer, & G. Theissen, "Catching a 'Hopeful Monster': Shepherd's Purse (Capsella Bursa-pastoris) as a Model System to Study the Evolution of Flower Development," Journal of Experimental Botany 57(13), 2006 p3531-3542.
Hoffmeyer (J.) & C. Emmeche, "Code-Duality and the Semiotics of Nature," J. Biosemiotics (1), 2005, p37-64.
Horgan (John), "The Consciousness Conundrum," IEEE Spectrum Online, 6/08.
Hunter (Cornelius), Darwin's Proof, 2003, p6 0.
Jenuwein (Thomas) & C. David Allis, "Translating the Histone Code," Science: 273, 8/10/01, p1074-1080.
Jimenez-Montano (Miguel), "Applications of Hyper Genetic Code to Bioinformatics," J. Biol. Sys.: 12, 2004, p5-20.
Johnson (D.), D. Lilja & J. Riedl, "A Distributed Hardware Mechanism for Process Synchronization on Shared-bus Multi-

processors," Int. Conf. on Parallel Proc.:II, 8/94, p268-275.
Johnson (D.), D. Lilja & J. Riedl, "A Circulating Active Barrier Synchronization Mechanism," International Conference on Parallel Processing:I, 8/95, p202-209.
Johnson (D.), D. Lilja, J. Riedl,& J. Anderson, "Low-Cost, High-Performance Barrier Synchronization on Networks of Workstations," Jour Par & Distributed Computing, 2/97B, p131-137.
Johnson (Donald), Exploring Fine-Grained Process Interaction in Multiprocessor Systems, University of Minncsota Thesis, 1997T.
Johnson (Donald), "Data and Information: Effect of Bioinformatics on Traditional Biology," Int Conf on Bioinformatics (Poster), 12/04.
Johnson (D.), D. Lilja & J. Riedl, "Circulating Shared-Registers for Multiprocessor Systems," Jour. Of Systems Arch., 6/3/05, p152-168.
Johnson (Donald), Probability's Nature and Nature's Probability: A Call to Scientific Integrity, 2009I.
Johnson (Donald), Probability's Nature and Nature's Probability - Lite: A Sequel for Non-Scientists and a Clarion Call to Scientific Integrity, 2009L.
Joyce (Gerald), "Nucleic Acid Enzymes: Playing with a Fuller Deck," PNAS: 95 (11), 5/26/98, p5845-5847.
Joyce (Gerald) & Leslie Orgel, "Prospects for Understanding the Origin of the RNA World," in The RNAWorld, 2nd ed.,1999, p49–77.
Kalinsky (K. D.), 2/19/08, http://www.newscholars.com/papers/ID%20Web%20Article.pdf.
Kauffman (Louis), CYBCON discusstion group 18, 9/20/07, p15.
Keith (Arthur), Evolution and Ethics, 1947, p230.
Kemsley (Jyllian), "Prebiotic Comet Collision Chemistry," C&EN 88 (13), 3/26/10, p21.
Kenyon (Dean), Biochemical Predestination, 1969.
Keplinger (B.L.), A.L. Rabetoy, & D.R. Cavener, "A Somatic Reproductive Organ Enhancer Complex Activates Expression in Both the Developing and the Mature Drosophila Reproductive Tract," Developmental Biology: 180, 1996, p311-323.
Kim (Y.), M, Coppey, R. Grossman, L. Ajuria, G. Jiménez, Z. Paroush, S. Shvartsman, "MAPK Substrate Competition Integrates Patterning Signals in the Drosophila Embryo," Current Biology: 20, 3/9/10, p1-6.
Kinoshita (S.), S. Kageyama, K. Iba, Y. Yamada, & H. Okada,, "Utilization of a Cyclic Dimer and Linear Oligomers of

e-aminocaproic Acid by Achromobacter Guttatus," Agricultural & Biological Chemistry (39), 6/75, p1219-23.

Kohler (J.), S. Schafer-Preuss, & D. Buttgereit, "Related Enhancers in the Intron of the Beta1 Tubulin Gene of Drosophila Melanogaster are Essential for Maternal and CNS-specific Expression During Embryogenesis," Nucleic Acids Research: 24, 1996, p2543-2550.

Kondo (T.), S. Plaza, J. Zanet, E. Benrabah, P. Valenti, Y. Hashimoto, S. Kobayashi, F. Payre, Y. Kageyama, "Small Peptides Switch the Transcriptional Activity of Shavenbaby During Drosophila Embryogenesis," Science 329, 7/16/10, p336-339.

Koonin (Eugene) & Artem Novozhilov, "Origin and Evolution of the Genetic Code: The Universal Enigma," arXiv:0807.4749, 7/08.

Koonin (Eugene), "Darwinian Evolution in the Light of Genomics," Nucleic Acids Research 37(4), 2/12/09, p1011–1034.

Kozmik (Z.), J. Ruzickova, K. Jonasova, Y. Matsumoto, P. Vopalensky, I. Kozmikova, H. Strnad, S. Kawamura, J. Piatigorsky, V. Paces, & C. Vlcek, "Assembly of the Cnidarian Camera-Type Eye from Vertebrate-Like Components," PNAS: 105 (26), 7/1/08, p8989–8993.

Knights (C.), J. Catania, S. Di Giovanni, & S. Muratoglu, "Distinct P53 Acetylation Cassettes Differentially Influence Gene-expression Patterns and Cell Fate," Jour of Cell Biology: 173, 2006, p553-544.

Kruger (K.), P. Grabowski, A. Zaug, J. Sands, D. Gottschling, & T. Cech, "Self-Splicing RNA: Autoexcision and Autocyclization of the Ribosomal RNA Intervening Sequence of Tetrahymena," Cell 31(1), 11/82, p147-157.

Kull (Kalevi), "A Brief History of Biosemiotics," in Biosemiotics: Information, Codes and Signs in Living Systems, 2007, p2.

Lad (Chetan), Nicholas Williams, & Richard Wolfenden, "The Rate of Yydrolysis of Phosphomonoester Dianions and the Exceptional Catalytic Proficiencies of Protein and Inositol Phosphatases," PNAS: 100 (10), 5/13/03, p5607-5610.

Lane (N.), J. Allen, & W. Martin, "How Did LUCA Make a Living? Chemiosmosis in the Origin of Life," Bioessays, 1/27/10, 10.1002/bies.200900131.

Lenski (Richard), Charles Ofria, Robert T. Pennock, & Christoph Adami, "The Evolutionary Origin of Complex Features," Nature: 423, 5/8/03, p139-44.

Le Page (Michael), "Genome at 10: A Dizzying Journey into Complexity," New Scientist, 6/16/10.

Lester (Lane), James C. Hefley, Human Cloning: Playing God or Scientific Progress?, 1998.
Levi-Setti (R.), Trilobites: A Photographic Atlas (2nd edition), 1993.
Lewin (Roger), "Evolutionary Theory under Fire," Science: 210, 1980, p883.
Lewis (Ricki), Bruce Parker, Douglas Gaffin, Marielle Hoefnagels, Life, 2006, Sec 13.5.
Lewontin (Richard), "Billions and Billions of Demons," in The NY Review of Books, 1/9/97.
Lieberman (Judy), "Master of the Cell," The Scientist, 4/1/10, p42.
Linton (Alan), "Scant Search for the Maker," The Times Higher Education Supplement, 4/ 20/01, Book Section, p29.
Lloyd (Seth), "Computational capacity of the Universe," Phys. Rev. Lett.: 88, 5/24/02, 4 pgs.
Lovgren (Stefan), "Computer Made from DNA and Enzymes," National Geographic, 2/24/03, news.nationalgeographic.com/news /2003/02/0224_030224_DNAcomputer.html
Ludwig (Mark), Computer Viruses, Artificial Life and Evolution, 1993.
Luisi (P.), "The Problem of Macromolecular Sequences: the Forgotten Stumbling Block," Origins of Life and Evolution of the Biosphere 37, 4–5/2007, p363–365.
Luskin (Casey), "Does Challenging Darwin Create Constitutional Jeopardy? A Comprehensive Survey of Case Law Regarding the Teaching of Biological Origins," Hamline University Law Review: 32(1), 2009, p1-64 (link: www.discovery.org/a/11291).
MacDónaill (Dónall), "Digital Parity and the Composition of the Nucleotide Alphabet," IEEE Engineering in Medicine and Biology, 1-2/06, p54-61.
Macnab (Robert), CRC Crit. Rev. Biochem: 5, 12/78, p333.
Maddox (John), What Remains to Be Discovered: Mapping the Secrets of the Universe, the Origins of Life, and the Future of the Human Race, 1998, p252.
Madrigal (Alexis), "Wired Science Reveals Secret Codes in Craig Venter's Artificial Genome," 1/28/08, blog.wired.com/wiredscience/2008/01/venter-institut.html.
Makalowski (Wojciech), "Not Junk After All.," Science, 5/23/03.
Mansy (Sheref), "Membrane Transport in Primitive Cells," Cold Spring Harb Perspect Biol, 4/21/10, doi: 10.1101
Margulis (Lynn) & Dorion Sagan, Acquiring Genomes: A Theory of

the Origins of the Species, 2003, p29.
Mattick (John), “Challenging the Dogma: The Hidden Layer of Non-Protein-Coding RNAs in Complex Organisms,” BioEssays: 25, 10/03, p930–939.
May (E.), M. Vouk, D. Bitzer, & D, Rosnick, “An Error-Correcting Code Framework for Genetic Sequence Analysis,” J. Frank Inst.: 341, 1-3/04, p89-109.
Mazur (Suzan), The Altenberg 16: An Exposé of the Evolution Industry, 2010.
McCormick (T.) & R.Fortey, “Independent Testing of a Paleobiological Hypothesis: the Optical Design of Two Ordovician Pelagic Trilobites Reveals Their Relative Paleobathymetry,” Paleobiology: 24 (2), 1998, p235–253.
McDonald (John), “The Molecular Basis of Adaptation: A Critical Review of Relevant Ideas and Observations,” Annual Review of Ecology and Systematics: 14, 1983, p77–102.
McIntosh (Andy), “Entropy, Free Energy and Information in Living Systems,” International Journal of Design & Nature and Ecodynamics 4 (4), 2009, p351-385.
Meyer (Stephen), Marcus Ross, Paul Nelson, & Paul Chien, “The Cambrian Explosion: Biology’s Big Bang,” Darwinism, Design and Public Education, 2003.
Meyer (Stephen), Signature in the Cell, 2009.
Miller (S. L.), “Production of Amino Acids Under Possible Primitive Earth Conditions,” Science: 117, 1953, p528.
Mims (Forrest), Rejected Letter, Science, www.forrestmims.org/publications.html,12/94.
Minnich (Scott) & Stephen C. Meyer, “Genetic Analysis of Coordinate Flagellar and Type III Regulatory Circuits in Pathogenic Bacteria,” Proceedings Second International Conf. on Design & Nature, 9/04.
Morowitz (Harold), Energy Flow in Biology, 1979, p99.
Morris (Conway), “Evolution: Bringing Molecules into the Fold,” Cell: 100, 1/7/00, p1-11.
Morrish (T.), N. Gilbert, J. Meyers, B. Vincent, T. Stamato, G. Taccioli, M. Batzer, & J. Moran, “DNA Repair Mediated by Endonuclease-Independent LINE-1 Retrotransposition,” Nature Genetics: 31, 2002, p159-165.
Murray (Daniel) & Scott W. Teare, “Probability of a Tossed Coin Landing on Edge,” Phys. Rev. E 48,10/93, p2547-2552.
Myers (PZ), “Junk DNA is Still Junk,” 5/19/10,

scienceblogs.com/pharyngula/2010/05/junk_dna_is_still_junk.php
Nakabachi (A.), A. Yamashita, H. Toh, H. Ishikawa, H. Dunbar, N. Moran, & M. Hattori, “The 160-Kilobase Genome of the Bacterial Endosymbiont Carsonella,” Science: 314 (5797), 10/13/06, p267.
Namba (Keiichi), “Self-Assembly of Bacterial Flagella,” 2002 Annual Meeting of the American Crystallographic Association, San Antonio, TX, www.aip.org/mgr/png/2002/174.htm.
New Scientist cover story on “The 10 Biggest Mysteries of Life,” 9/4/2004.
Nobel-1989,nobelprize.org/nobel_prizes /chemistry/laureates/1989/press.html
Nobel-2004, nobelprize.org/nobel_prizes /chemistry/laureates/2004/illpres/5_proteins.html
Ofria (C.), C. Adami, & T. Collier, “Selective Pressures on Genomes in Molecular Evolution,” Journal of Theoretical Biology.: 222, 2003, p477-483.
Ohno (Susumu), "So Much 'Junk' DNA in Our Genome," Brook Haven Symposia in Biology: 23, 1972, p366-370.
OOLprize, Origin of Life Prize, www.us.net/life/
Orgel (Leslie), “The Origin of Life on the Earth,” Sci Am.. 271, 10/94, p76-83.
Orgel (Leslie), “Prebiotic Chemistry and the Origin of the RNA World,” Crit.l Rev. in Biochemistry and Molecular Biology: 39, 2004, p99–123.
Orgel (Leslie), “The Implausibility of Metabolic Cycles on the Prebiotic Earth,” PLoS Biol 6(1), 2008, e18.
Orr (H. A.) & Jerry Coyne, “The Genetics of Adaptation: a Reassessment,” Am Nat, 1992, p726.
Pandolfi (Pier), quoted in “Language of RNA Decoded: Study Reveals New Function for Pseudogenes and Noncoding RNAs,” Science Daily, 6/24/10.
Pattee (Howard), “Quantum Mechanics, Heredity and the Origin of Life,” J. Theor. Biol. (17), 12/17/67, p410-20.
Pauling (Linus), The Linus Pauling Institute Website, lpi.oregonstate.edu/lpbio/lpbio2.html
Pearson (Helen), “'Junk' DNA Reveals Vital Role,” Nature, 5/3/04.
Pennock (Robert), “Does Design Require a Designer,” God, Design & Nature conference, Oxford University, Oxford UK, 7/10/08.
People94, www.ncbi.nlm.nih.gov/pubmed/12288594
Perkins (D. O.), C. Jeffries, & P. Sullivan, “Expanding the ‘Central

Dogma': the Regulatory Role of Nonprotein Coding Genes and Implications for the Genetic Liability to Schizophrenia," Molecular Psychiatry: 10, 2005, p69–78.

Pigliucci (Massimo) & Gerd Müller, "Elements of an Extended Evolutionary Synthesis," in Evolution—the Extended Synthesis, 2010, p3-18.

Polanyi (Michael), Quote originally at Michael Polanyi Center site, now at nostalgia.wikipedia.org/wiki/Michael_Polanyi

Poliseno (L.), L. Salmena, J. Zhang, B. Carver, W. Haveman, & P. Pandolfi, "A Coding-Independent Function of Gene and Pseudogene mRNAs Regulates Tumour Biology," Nature, 6/23/10.

Popper (Karl), "Science as Falsification," Conjectures and Refutations, 1963, p33-39.

Powell (Alvin), "NYU Chemist Robert Shapiro Decries RNA-First Possibility," Harvard News Office, 10/23/08.

Prigogine (I.), N. Gregair, A. Babbyabtz, "Thermodynamics of Evolution," Physics Today: 25, 1972, p23-28.

Provine (Will), "No Free Will," in Catching Up with the Vision, Ed. By Margaret W. Rossiter, 1999, pS123.

PSSI, Physicians and Surgeons Who Dissent from Darwinism, http://www.pssiinternational.com/list.pdf

Raichle (Marcus) & Debra Gusnard, "Appraising the Brain's Energy Budget," PNAS: 99 (16), 8/6/02, p10237-10239.

Ray (C.), "DNA; Junk or Not," The New York Times, 3/4/03.

Ray (T.), "Evolution, Ecology, and Optimization of Digital Organisms," 1992, www.htp.atr.co.jp/~ray/pubs/tierra/tierrahtml.html

Ridley (Mark), The Cooperative Gene: How Mendel's Demon Explains the Evolution of Complex Beings, 2001, p111.

Ridley (Mat), Genome: Autobiography of a Species in 23 Chapters, 1999, p21-22.

Rockefeller University, "Scientists Crash Test DNA's Replication Machinery," ScienceDaily, 3/1/10, http://www.sciencedaily.com/releases/2010/02/100227212116.htm

Rojas (Raul), "How to Make Zuse's Z3 a Universal Computer," IEEE Annals of the History of Computing: 20(3), 1998, p51–54.

Ruse, "Saving Darwinism from the Darwinians," National Post, 5/13/00, pB.

Ruse (Michael) & E. O. Wilson, "The Evolution of Ethics," in Religion and the Natural Sciences: The Range of Engagement, 1991.

Sagan (Carl), "Can We Know the Universe?," in Broca's Brain, 1979, p13-18.
Sagan (Carl), Contact: a novel, 1985.
Sagan (Carl) & Ann Druyan, Shadows of Forgotten Ancestors, 1992, p128.
Sagan (Carl), "Life," Encyclopaedia Britannica: 22, 1997, p964-981.
Salk Institute, "Nuclear Pore Complexes Harbor New Class of Gene Regulators," Science Daily, 2/11/10, http://www.sciencedaily.com/releases/2010/02/100204144424.htm
Sarkar (Sahotra), "Biological Information: A Skeptical Look at Some Central Dogmas of Molecular Biology," in The Philosophy and History of Molecular Biology: New Perspectives, 1996, p191.
Schecter (Julie), "How Did Sex Come About?," Bioscience: 34, 1984, p680.
Schnorrer (F.), C. Schönbauer, C. Langer, G. Dietzl, M. Novatchkova, K. Schernhuber, M. Fellner, A. Azaryan, M. Radolf, A. Stark, K. Keleman, & B. Dickson, "Systematic Genetic Analysis of Muscle Morphogenesis and Function in Drosophila," Nature, 3/11/10.
Segre (Daniel), Dafna Ben-Eli, & Doron Lancet, "Compositional Genomes: Prebiotic Information Transfer in Mutually Catalytic Noncovalent Assemblies," PNAS 97 (8), 4/11/00, p4112–4117.
Serrano (Luis), quoted in "First-Ever Blueprint of 'Minimal Cell" Is More Complex Than Expected," Science, 11/27/09.
SETI Website, setiathome.ssl.berkeley.edu/
Shannon (Claude), "A Mathematical Theory of Communication," Bell System Technical Journal: 27, July & October, 1948, p379-423 & 623-656.
Shannon (Claude), "Prediction and Entropy of Printed English," The Bell System Tech. Journal: 30, 1950, p50-64.
Shapiro (Robert), "A Simpler Origin for Life," Scientific American, 2/12/07.
Shapiro (James), "Mobile DNA and Evolution in the 21st Century," Mobile DNA 2010 1:4.
Shenhav (Barak), Aia Oz, & Doron Lancet, "Coevolution of Compositional Protocells and Their Environment," Phil. Trans. R. Soc. B 362, 2007, p1813-1819.
Simons (Andrew), "The Continuity of Microevolution and Macroevolution," Journal of Evolutionary Biology: 15, 2002, p688-701.
Skell (Philip), "Why Do We Invoke Darwin? Evolutionary Theory

Contributes Little to Experimental Biology," The Scientist, 8/29/05.
Smith (John Maynard), Evolutionary Genetics, 1989, p61.
Smith (Wolfgang), "The Universe is Ultimately to be Explained in Terms of a Metacosmic Reality," in Cosmos, Bios, Theos, 1992, p113.
Spetner (Lee), Not By Chance, 1997.
Spies (Maria), "Researchers Probe a DNA Repair Enzyme," Bio-Medicine, 2/18/08, www.bio-medicine.org/biology-news-1 /Researchers-probe-a-DNA-repair-enzyme-2257-1/
Stein (Lincoln), "Human Genome: End of the Beginning," Nature, 10/21/04, p431.
Steinhaus (H.), Mathematical Snapshots, 3rd ed., 1999, p202.
Sternberg (R. von) & J. Sharpiro, "How Repeated Retroelements Format Genome Function," Cytogenetic and Genome Research: 110, 2005, p108-116.
Supreme Court Decision, Torcaso v. Watkins (367 U.S. 488), 1961.
Supreme Court Decision, US v. Seeger, 380 U.S. 163, 1965
Supreme Court Decision, Gillette v. U.S., 401 U.S. 437, 450, 1971.
Supreme Court Decision, Harris v. McRae, 448 U.S. 297, 1980.
Sutherland (John), "Ribonucleotides," Cold Spring Harb Perspect Biol, 3/10/10, 2:a005439.
Swee-Eng (A. W.), "The Origin of Life: A Critique of Current Scientific Models," CEN Tech. J.: 10-3, 1996, p300-314.
Szathmary (Eors), "The Origin of Replicators and Reproducers," Phil. Trans. R. Soc. B 361, 2006, p1761–1776.
Szöllsi (Gergely), Imre Derényi, & Tibor Vellai, "The Maintenance of Sex in Bacteria Is Ensured by Its Potential to Reload Genes," Genetics: 174, 12/06, p2173-2180.
Szostak (Jack), "Functional Information: Molecular Messages," Nature: 423, 6/ 12/ 2003, p689.
Tenaillon (Olivier), Hervé Le Nagard, Bernard Godelle, and François Taddei, "Mutators and Sex in Bacteria: Conflict Between Adaptive Strategies," PNAS: 97(19), 9/12/00, p10465-10470.
Tennant (Richard), The American Sign Language Handshape Dictionary, 1998.
Thomas (A.T.), "Developmental Palaeobiology of Trilobite Eyes and Its Evolutionary Significance," Earth-Science Reviews: 71 (1-2), 6/05, p77-93.
Trefil (James), Harold Morowitz, Eric Smith, "The Origin of Life," American Scientist 97(2), 3-4/09, p206-210.

Trevors (J. T.) & D. L. Abel, "Chance and Necessity Do Not Explain the Origin of Life," Cell Biology International: 28, 2004, p729-739.
Truman (Royal), "Evaluation of Neo-Darwinian Theory Using the Avida Platform," PCID 3.1.1, 11/04.
Tuteja (N.) & R. Tuteja, "Unraveling DNA Helicases, Motif, Structure, Mechanism and Function," Eur J Biochem: 271 (10), 2004, p1849–63.
Valentine (J. W.), et al., "Fossils, Molecules, and Embryos: New Perspectives on the Cambrian Explosion," Development: 126, 1999, p851-59.
Vasas (V.), E.Szathmáry, & M.Santos, "Lack of Evolvability in Self-sustaining Autocatalytic Networks Constraints Metabolism-first Scenarios for the Origin of Life," PNAS, 2/11/10, www.pnas.org/content/107/4/1470.
Veeramachaneni (V.), W. Makalowski1, M. Galdzicki, R. Sood, & I. Makalowska, "Mammalian Overlapping Genes: The Comparative Perspective," Genome Res.: 14, 2004, p280-286.
Venter (Craig), Interview "A Bug to Save the Planet," Newsweek, 6/16/08, p40.
Voet (D.) & J. Voet, Biochemistry, 1995, p1138.
Voie (Albert), "Biological Function and the Genetic Code are Interdependent," Chaos, Solitons and Fractals: 28(4), 2006, p1000-1004.
Wald (George), "Innovation and Biology," Scientific American:199, 9/58, p100.
Watson (James) & Francis Crick, "Molecular structure of Nucleic Acids," Nature: 171, 1953, p737–738.
Webster's Dictionary for Everyday Use, 1987.
Webster's Third New International Dictionary of the English Language, unabridged, 1993.
Weikart (Richard), From Darwin to Hitler: Evolutionary Ethics, Eugenics, and Racism in Germany, 2004.
Wells (Jonathan), Icons of Evolution, 2000.
Wiedersheim (Robert), The Structure of Man: An Index to His Past History, 1893.
Wilder-Smith (A.E.), The Scientific Alternative to Neo-Darwinian Evolutionary Theory, 1987, p73.
Williams (Bryony), Claudio Slamovits, Nicola Patron, Naomi Fast, & Patrick Keeling, "A High Frequency of Overlapping Gene Expression in Compacted Eukaryotic Genomes," PNAS: 102 no. 31,

8/2/05, p10936-10941.
Williams (George), "A Package of Information," in The Third Culture: Beyond the Scientific Revolution, 1995, p42-43.
Wilson.(David), "Atheism as a Stealth Religion," Huffington Post, 12/14/07.
Wilson (Edgar), An Introduction to Scientific Research, 1990.
Winther (Rasmus), "Systemic Darwinism," PNAS, 8/19/08, p11833-8.
Witzany (Günther), Natural Genetic Engineering and Natural Genome Editing (Proceedings), 12/09, 276 pgs
Wi-Web, www.wisconsincountyforests.com/qa-forst.htm
Woese (Carl), The Genetic Code, the Molecular Basis for Genetic Expression, 1967.
Woese (Carl), "The Universal Ancestor," Proceedings of the National Academy of Sciences USA: 95, 6/98, p6854-9859.
Woese (Carl), quoted in "Horizontal and Vertical: the Evolution of Evolution," New Scientist, 1/28/10.
Wolfenden (Richard), in "Without Enzyme Catalyst, Slowest Known Biological Reaction Takes 1 Trillion Years," 2003, www.unc.edu/news/archives/may03/enzyme050503.html.
Yan (F.), A. Bhardwaj,& M. Gerstein, "Comparing Genomes to Computer Operating Systems in Terms of the Topology and Evolution of Their Regulatory Control Networks," PNAS, 5/3/10, 6 pgs.
Yeh (Edward), quoted in "Key Step for Regulating Embryonic Development Discovered," Science Daily, 4/22/10.
Yockey (Hubert), "A Calculation of the Probability of Spontaneous Biogenesis by Information Theory," J. Theor. Biol., 1977, p377–398.
Yockey (Hubert), Information Theory and Molecular Biology, 1992.
Yockey (Hubert), Information Theory, Evolution, and the Origin of Life, 2005.
Yokoyama (Shozo), in "Evolution of a Fish Vision Protein," Physics Today, 9/2/08, http://blogs.physicstoday.org/update /2008/09/evolution_of_a_fish_vision_pro.html
Zewail (Ahmed), Nobel Prize in Chemistry, 1999.
Zimmer (Carl), "Testing Darwin," Discover Magazine, 2/5/05, Cover.
Ziv (Jacob) & Abraham Lempel, "Compression of Individual Sequences Via Variable-Rate Coding," IEEE Transactions on Information Theory, 9/78, p530-536.
Zuckerkandl (Emile), "Neutral and Nonneutral Mutations: The Creative Mix-Evolution of Complexity in Gene Interaction Systems," Journal of Molecular Evolution: 44, 1997, p53.

Appendix A: Logarithms, Probability, and Other Math

Logarithms can be used to calculate the exponent for any base: $\log_b(b^x) = x$. Examples include: $\log_{10}(1000) = 3$, $\log_{10}(1/1000) = -3$, $\log_2(16) = 4$ ($2^4 = 16$), $\log_2(.25) = -2$ ($1/4 = 2^{-2}$), $\log_{10}(5) \sim 0.69897$, $\log_{10}(50) \sim 1.69897$, $\log_{10}(6.02 \times 10^{23}) = \log_{10}(10^{23.77975}) = 23.77975$, $\log_{10}(1) = \log_2(1) = \log_e(1) = 0$, $\log_e(10) \sim 2.3025$ ($\log_e$ is known as the natural logarithm, which is the basis of many math expressions).

Base conversion: $\log_d(x) = \log_b(x)/\log_b(d)$

If the probability of an outcome is P, the number of trials (n) for that outcome to become probable is $n = \log_2(0.5)/\log_2(1-P) = -1/\log_2(1-P)$.

Taylor's expansion of $\log_e(1 + x) = -\sum_{i=1}^{\infty} (x^i/i)(-1)^{(i+1)}$ for $-1 < x \leq 1$

Bernoulli numbers (B_i) are a sequence of signed rational numbers that can be defined by the identity:

$$x/(e^x - 1) \equiv -\sum_{i=0}^{\infty} B_i x^i/i!$$

Ada Lovelace, wrote a solution to this in 1843 (the first computer program).

The selective value (S) of an organism's genome can express the relative increase (or decrease, if negative) in reproductive probability, such that a mutated organism with S of 0.1% would have 0.1% more offspring than the average organism. The probability of that mutant strain surviving, assuming equal mating probability, is $2S/(1 - e^{-2SN})$ [Spe97p80], where N is the total population size.

Appendix B: Comparison of Computer Disk Drive and DNA

Excerpts and summary information is from the open access paper "A comparative approach for the investigation of biological information processing: An examination of the structure and function of computer hard drives and DNA" by David J D'Onofrio and Gary An [D'On10]. Perhaps reverse engineering of the cybernetic complex computing systems of life to determine characteristics of the native language(s) and the operating system(s) in the enzymes, ribosomes, and tRNA may yield breakthroughs in miniaturization and performance.

Abstract

Background: The robust storage, updating and utilization of information are necessary for the maintenance and perpetuation of dynamic systems. These systems can exist as constructs of metal-oxide semiconductors and silicon, as in a digital computer, or in the "wetware" of organic compounds, proteins and nucleic acids that make up biological organisms. We propose that there are essential functional properties of centralized information processing systems; for digital computers these properties reside in the computer's hard drive, and for eukaryotic cells they are manifest in the DNA and associated structures.

Methods: Presented herein is a descriptive framework that compares DNA and its associated proteins and subnuclear structure with the structure and function of the computer hard drive. We identify four essential properties of information for a centralized storage and processing system: (1) orthogonal uniqueness, (2) low level formatting, (3) high level formatting and (4) translation of stored to usable form. The corresponding aspects of the DNA complex and a computer hard drive are categorized using this classification. This is intended to demonstrate a functional equivalence between the components of the two systems, and thus the systems themselves.

Results: Both the DNA complex and the computer hard drive contain components that fulfill the essential properties of a centralized information storage and processing system. The functional equivalence of these components provides insight into both the design process of engineered systems and the evolved solutions addressing similar system requirements. However, there are points where the comparison breaks down, particularly when there are externally imposed information-organizing structures on the computer hard drive. A specific example of this is the imposition of the File Allocation Table (FAT) during high level formatting of the computer hard drive and the subsequent loading of an operating system (OS). Biological systems do not have an external source for a map of their stored information or for an operational instruction set; rather, they must contain an organizational template conserved within their intra-nuclear architecture that "manipulates" the laws of chemistry and physics into a highly robust instruction set. We propose that the epigenetic structure of the intra-nuclear environment and the non-coding RNA may play the roles of a Biological File Allocation Table (BFAT) and biological operating system (Bio-OS) in eukaryotic cells.

Conclusions: The comparison of functional and structural characteristics of the DNA complex and the computer hard drive leads to a new descriptive paradigm that identifies the DNA as a dynamic storage system of biological information. This system is embodied in an autonomous operating system that inductively follows organizational structures, data hierarchy and executable operations that are well understood in the computer science industry. Characterizing the "DNA hard drive" in this fashion can lead to insights arising from discrepancies in the descriptive framework, particularly with respect to positing the role of epigenetic processes in an information-processing context. Further expansions arising from this comparison include the view of cells as parallel computing machines and a new approach towards characterizing cellular control systems.

Appendix C: Life Details and Origin Speculations

Amino acids are the building-blocks of life. Each is an organic molecule that has a carboxylic acid ($-CO_2^-$) and an amine ($-NH_3^+$) group attached to the same carbon atom, which means there is α-linkage. Life uses only α-linkage amino acids that are levorotatory (l) – "left-handed" chirality, rotating polarized light counter-clockwise. **Chirality** divides two compounds having the same chemical formula and structure, but that are non-superimposable mirror images of each other, into "l" and "d" versions called enantiomers. Human hands, for example, are mirror-images, but can't be superimposed on each other, and are analogies of right-handed and left-handed enantiomers. In chemistry, both versions are equivalent and are always generated as a racemic (both l and d) mixture, but in biochemistry (life), only one form is useful.

The second carbon (next to the α-linkage carbon) of an amino acid has another chemical side-group (or side chain) attached. There are 20 main amino acids for life, each with a standard 3-letter and a 1-letter abbreviation. The 1-letter abbreviation – any letter except BJOUXZ – has great utility when using the NCBI Entrez [Entr] protein database, has been used to create "watermarks" (Appendix E topic) for artificial genes [Mad08], and are used in this book for amino acid identification. The amino acid names won't normally be used in this rudimentary coverage.

A **peptide** bond results during a dehydration reaction in which the carboxylic group of one amino acid reacts (assisted by an enzyme that makes the reaction feasible) with the amino group of another to remove a water (H_2O) molecule (like an acid-base reaction does). Multiple such reactions link amino acids together into **polypeptide** chains, including **proteins**. Sometimes unusable proteins are constructed. An excerpt from the description of the 2004 Nobel Prize in Chemistry states, *"Surprisingly many of the proteins created in the cell are faulty from the start. They must be broken down and rebuilt since they can damage the organism... When the proteins have been hacked to pieces, the cell can use their amino acids to synthesize other proteins. When protein degradation does not function correctly, we can become ill"* [Nob04].

DNA is Chemically a long polymer of nucleotides, each **nucleotide** being a deoxyribose sugar molecule, one phosphate group, and one **base**. Nucleotides are joined by ester bonds so that the sugars and phosphates form the DNA backbone, with the bases sticking out to form hydrogen bonds with a second DNA strand to form a double helix [Wat53]. The four bases used in DNA are adenine (A), cytosine (C), guanine (G) and thymine (T), with base pairing (such as between strands of the helix) always being G-C or A-T. **Ribonucleic acid (RNA)**, with ribose replacing deoxyribose, is normally single-stranded, except when in a hybrid helix with DNA during RNA construction.

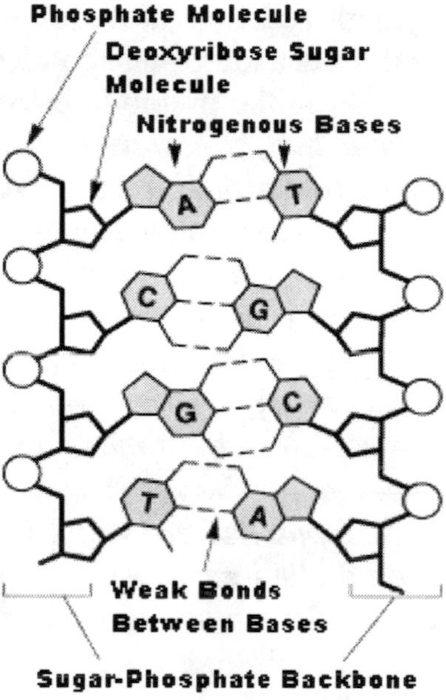

The RNA world's *"Molecular Biologists Dream"* [Joy99] for its origin *"can be strung together from optimistic extrapolations of the various achievements of prebiotic chemistry and directed RNA evolution.* [One could reasonably wonder what is meant by "directed."] ... *First we suppose that nucleoside bases and sugars were formed by prebiotic reactions on the primitive Earth and/or brought to the Earth in meteorites, comets, etc. Next, nucleotides were formed from prebiotic bases, sugars, and inorganic phosphates or polyphosphates, and they accumulated in an adequately pure state* [how – what would cause the purification?] *in some special little 'pool.' A mineral catalyst at the bottom of the pool—for example, montmorillonite—then catalyzed the formation of long single-stranded polynucleotides, some of which were then converted to complementary double strands by template-directed synthesis* [template cause for this prescriptive information?]. *In this way a library of double-stranded RNAs accumulated on the primitive Earth. We suppose that among the double-stranded RNAs there was at least one that on melting yielded a (singlestranded) ribozyme capable of copying itself and its complement. Copying the complement would then have produced a second ribozyme molecule, and then repeated copying of the ribozyme and its complement would... lead to an exponentially growing population. In this scenario this is where natural selection takes over"* [Org04]. One could wonder how the "adequately pure" materials or the "library" of RNA formed, or how a mineral crystal could serve as a template for producing information. Any information in a crystal structure would have to be in irregularities, as the regular structure contains almost no information.

Orgel points out several remaining problems including the nonenzymatic (enzymes are required for known life) synthesis, polymerization, and replication of nucleotides to produce RNA capable of exponential growth in the prebiotic environment, and that *"difficulties remain so severe that alternatives to the de novo appearance of RNA on the primitive Earth deserve serious consideration"* [Org04]. *"But where the first RNA came from is a mystery; it's hard to see how the chemicals on early Earth could have combined to form the complicated nucleotides that make up RNA"* [Dav00]. *"What is essential, therefore, is a reasonably detailed description, hopefully supported by experimental evidence, of how an evolvable family of cycles might operate. The scheme should not make unreasonable demands on the efficiency and specificity of the various external and internally generated catalysts that are supposed to be involved. Without such a description, acceptance of the possibility of complex nonenzymatic cyclic organizations that are capable of evolution can only be based on faith, a notoriously dangerous route to scientific progress"* [Org08].

"Replicators are fundamental to the origin of life and evolvability. Their survival depends on the accuracy of replication and the efficiency of growth relative to spontaneous decay" [Sza06]. As De Duve (Nobel Prize-winning biochemist) observes, *"The problem is not as simple as might appear at first glance. Attempts at engineering – with considerably more foresight and technical support than the prebiotic world could have enjoyed – an RNA molecule capable of catalyzing RNA replication have failed so far"* [DeD95]. *"The prebiotic synthesis of nucleotides in a sufficiently pure state to support RNA synthesis cannot be achieved using presently known chemistry"* [Org04]. *"Ribosome creation requires many RNA-modification enzymes that are still unknown"* [Bar08L]. *"There is no evidence that transcription or RNA replication involve ribozyme catalysis... One must recognize that, despite considerable progress, the problem of the origin of the RNA World is far from being solved"* [Org04].

"The reaction with purine nucleobases is low-yielding and the reaction with the canonical pyrimidine nucleobases does not work at all... difficulties in nucleobase ribosylation can be overcome with directing, blocking, and activating groups on the nucleobase and ribose... These molecular interventions are synthetically ingenious, but serve to emphasize the enormous difficulties that must be overcome if ribonucleosides are to be efficiently produced by nucleobase ribosylation under prebiotically plausible conditions. This impasse has led most people to abandon the idea that RNA might have assembled abiotically,

and has prompted a search for potential pre-RNA informational molecules.. Surely, it makes more sense to first find predisposed chemical routes to molecules of interest and then ask whether the sequence of conditions is geochemically plausible" [Sut10]. Meyer [Mey09p301-321] highlights many of the RNA world problems including the difficult synthesis, easy destruction, ineffectual catalysis, and the development of a coding system in the RNA. He also points out that designed experiments do not verify undirected processes.

Metabolism first is an alternative to genetics first. "*The essential problem is that in modern living systems, chemical reactions in cells are mediated by protein catalysts called enzymes. The information encoded in the nucleic acids DNA and RNA is required to make the proteins; yet the proteins are required to make the nucleic acids. Furthermore, both proteins and nucleic acids are large molecules consisting of strings of small component molecules whose synthesis is supervised by proteins and nucleic acids. We have two chickens, two eggs, and no answer to the old problem of which came first... The RNA molecule is too complex, requiring assembly first of the monomeric constituents of RNA, then assembly of strings of monomers into polymers. As a random event without a highly structured chemical context, this sequence has a forbiddingly low probability and the process lacks a plausible chemical explanation, despite considerable effort to supply one"* [Tre09].

"*The sudden appearance of a large self-copying molecule such as RNA was exceedingly improbable. Energy-driven networks of small molecules afford better odds as the initiators of life... inanimate nature has a bias toward the formation of molecules made of fewer rather than greater numbers of carbon atoms, and thus shows no partiality in favor of creating the building blocks of our kind of life*" [Sha07]. Shapiro believes metabolism first "*would require five things: some sort of a boundary to keep the ingredients together, such as a rock-bound compartment; a supply of energy; a coupling of the energy to a 'driver reaction'; a chemical network that would permit adaptation and evolution; and reproduction. Though such a model would have no single molecule holding the genetic information, like modern cells do, Shapiro said that wouldn't necessarily preclude replication"* [Pow08]. Many [Ben10, Kem10, Ber02, Ber06P, Coo01] believe that life's original chemical may have come from extraterrestrial sources.

Osmosis first scenario is based on thermodynamic constraints that *"make chemiosmosis strictly necessary for carbon and energy metabolism in all free-living chemotrophs, and presumably the first* *free-living cells too... One of the biggest stumbling blocks to the idea of 'chemiosmosis early' is the daunting complexity of the ATP synthase – a nanomachine comprising a rotary motor powered by a flow of protons through the membrane stalk, coupled to the rotating head that forms ATP from ADP and phosphate. The two main domains, the stalk and the rotating head, have no obvious homologues among other proteins"* [Lan10]. *"Model protocellular membranes allow for the passage of polar solutes and thus can potentially support cell-to functions without the aid of transport machinery...Thus far a robust and complete cycle including both nucleic acid and compartment replication has not been shown"* [Man10].

Simulations [She07] have been done on "composomes," packages of chemicals that have reaction cycles, a mechanism to harness energy, and reproduce by "bag division." *"Sets of compositional assemblies bear formal resemblance to quasi-species of biopolymers, providing a bridge between the 'genome first' and 'metabolism first' paradigms"* [Seg00]. *"The question of life's origin is in essence a problem of information transfer from a geochemical environment to a highly localized volume...* [and] *required the selection, concentration, and organization of specific organic molecules into successively more information-rich localized assemblages"* [Haz10].

"The unexpected levels of complexity revealed at the molecular level have further strained the concept of the random assembly of a self-replicating system" [Swe96]. *"Functionally effective proteins have a vanishingly small chance of arising spontaneously in a prebiotic environment"* [Jim04]. *"When discussing organic evolution the only point of agreement seems to be: 'It happened.' Thereafter, there is little consensus, which at first sight must seem rather odd"* [Mor00]. *"Proteins—and nucleic acids—are not simply polymers, but are co-polymers, and the kinetics and thermodynamics attending the synthesis of copolymers poses stringent constraints for the biogenesis and growth of specific sequences... there are no reliable methods described in the literature to make copolymers of amino acids or nucleotides under prebiotic conditions"* [Lui07]. *"All speculation on the origin of life on Earth by chance can not survive the first criterion of life: proteins are left-handed, sugars in DNA and RNA are right-handed... Unfortunately, the interpretations of the corpus of publications on the origin of life is false. Those experiments are based on a belief that life is just complicated chemistry and that the origin of life, if it could be found, is emergent from organic chemistry"* [Yoc05p119&147]. *"The likelihood*

of life having occurred through a chemical accident is, for all intents and purposes, zero. This does not mean that faith in a miraculous accident will not continue. But it does mean that those who believe it do so because they are philosophically committed to the notion that all that exists is matter and its motion. In other words, they do so for reasons of philosophy and not science" [Gan86].

Ignoring information to just consider chemistry, the probability of forming (allowing multiple configurations) a simple life-compatible protein of 150 amino acids is 10^{-164} [Mey09p212]. This is less than the probability of randomly selecting two specified atoms of the Universe. The calculated maximum polypeptide (proteins are polypeptides) expected in 10^9 years from a pool of pure, activated biological amino acids would only be 49 amino acids long [Yoc77].

Although some have said that the equilibrium calculations of Morowitz don't apply to the prebiotic environment, he has estimated that the probability for the chance formation of the smallest, simplest form of living organism known is one in $10^{340,000,000}$ [Mor79]. There is no way to know what the prebiotic environment may have been, and whether or not that environment was at equilibrium conditions, since most systems approach equilibrium exponentially.

Although some speculate that creation of life artificially is near [Bar08L], all approaches use existing life components for the attempted synthesis. *"We once thought that the cell, the basic unit of life, was a simple bag of protoplasm. Then we learned that each cell in any life form is a teeming micro-universe of compartments, structures, and chemical agents—and each human being has billions of cells"* [Les98p30-31]. *"Nothing in the modern synthesis* [Darwinism] *explains the most fundamental steps in early life: how evolution could have produced the genetic code and the basic genetic machinery used by all organisms, especially the enzymes and structures involved in translating genetic information into proteins. Most biologists, following Francis Crick, simply supposed that these were uninformative 'accidents of history.' That was a big mistake,"* says Microbiologist Carl Woese [Buc10].

Appendix D: Shannon Information Technical Details

Shannon binary information (often called entropy) may be calculated by: $\mathbf{H(S) = -\sum_{i=1}^{n} p_i \log_2(p_i)}$, where the source alphabet, $S = \{a_1, \ldots, a_n\}$, has discrete probability distribution $\{p_1, \ldots, p_n\}$ where p_i is the probability of the symbol a_i.

The probability (without thermodynamic considerations) of a N-long sequence of symbols from a finite alphabet is very close to 2^{-NH}, where NH is the total Shannon information, as opposed to the lower n^{-N} that is often assumed [Yoc05p28-9].

Given the probability vector, $\mathbf{p}_A$, of the elements of alphabet A in a source probability space $[\Omega, A, \mathbf{p}_A]$ and the probability vector, $\mathbf{p}_B$, of the elements of alphabet B in destination probability space $[\Omega, B, \mathbf{p}_B]$, then a unique mapping of the symbols of alphabet A onto the symbols of alphabet B is called a code.

Mutual entropy is a mathematical measure of the similarity between any two sequences one wishes to compare. Mutual entropy relates the input (x) and output (y) channels via: $\mathbf{I(B;A) = I(A;B) = H(x) - H(x|y)}$, where the conditional (x_i given y_i received) entropy is

$\mathbf{H(x|y) = -\Sigma_{ij} p_j\, p(i|j) \log_2 p(i|j)}$, $\mathbf{p_j = \sum_{i}^{n} p_i p(j|i)}$ (which relates the probability vector, $\mathbf{p}$, elements to those of the conditional probability matrix, $\mathbf{P}$), and $\mathbf{H(x) = -\sum_{i=1}^{n} p_i \log_2(p_i)}$ is the information entropy. The Shannon Channel Capacity is also the maximum mutual entropy.

For a transmitting system with fewer symbols in $[\Omega, A, \mathbf{p}_A]$ to pass information to $[\Omega, B, \mathbf{p}_B]$, the maximum mutual entropy would be exceeded.

Appendix E: Functional Information Technical Examples

Functional information has been quantified by extending Shannon uncertainty with a functionality variable to calculate Functional bits (Fits) of information. *"This explicitly incorporates empirical knowledge of metabolic function into the measure that is usually important for evaluating sequence complexity"* [Dur07]. Since genes can be thought of as information-processing subroutines, proteins can be analyzed in terms of the products of information interacting with laws of physics to advance our knowledge of the structure and functions of proteins. The patterns of functional information are examined for a protein family. The method proposed is based on mathematical and computational concepts.

The resulting equation is $\mathbf{H_f(t) = -\Sigma P(X_f(t)) \log P(X_f(t))}$, where x_f denotes the conditional variable of the given sequence data (X) on the described biological function f, which is an outcome of the variable (F), using the joint variable (X,F). This results in an exponential decrease in probability with a linear increase in functional sequence complexity (FSC). *"For example, 342-residue SecY has a FSC of 688 Fits, but the smaller 240-residue RecA actually has a larger FSC of 832 Fits. The Fit density (Fits/amino acid) is, therefore, lower in SecY than in RecA. This indicates that RecA is likely more functionally complex than SecY"*[Dur07]

*"**Complex emergent systems** of many interacting components, including complex biological systems, have the potential to perform quantifiable functions. Accordingly, we define 'functional information,' $I(E_x)$, as a measure of system complexity. For a given system and function, x (e.g., a folded RNA sequence that binds to GTP), and degree of function, E_x (e.g., the RNA–GTP binding energy), $I(E_x) = -\log_2[F(E_x)]$, where $F(E_x)$ is the fraction of all possible configurations of the system that possess a degree of function E_x... Function is thus the essence of complex systems. Accordingly, we focus on function in our operational definition of complexity. Therefore, although many previous investigators have explored aspects of biological systems in terms of information..., we adopt a different approach and explore information in terms of the function of a system (including biological systems)... In this formulation, functional information increases with degree of function, from zero for no function (or minimum function) to a maximum value corresponding to the number of bits necessary and sufficient to specify completely any configuration of that system"* [Haz07].

Formalism and physicality are distinguished by Kalinsky by the ability *"to produce effects requiring significant levels of functional information"* [Kal08p2]. He points out that natural selection is credited with discovering the nucleotides to code for thousands of proteins, each with a stable 3-D structure, including those used in extremely complex molecular machines and molecular computers. Rigor can be introduced into analyzing the process using principles of genetic and evolutionary algorithms, which require fitness functions to avoid a blind (e.g. random walk) search. A fitness function represents *"the requirements to adapt to. It forms the basis for selection, and thereby it facilitates improvements. More accurately, it defines what improvement means. From the problem-solving perspective, it represents the task to solve in the evolutionary context"* [Eib03p19]. The fitness function must contain at least as much functional information as the desired outcome or the information deficit must be made up for in a blind search. The fitness function for natural selection is unknown, but without such a function that contains sufficient functional information required to find each protein in an evolutionary search, a blind search results. If natural selection is responsible for the origin of a protein-coding gene, the amount of functional information in the natural fitness function (that initially would be in a non-life form for biogenesis) can be estimated by measuring the functional information required by a given protein. A specified level of functionality can become probable after R trials of a specified function with a single sampling probability, P_f, when: $0.5 = 1 - (1 - P_f)^R$, or $P_f = 1 - (1 - 0.5)^{1/R}$. If only one configuration meets the required function, the functional information that could occur by mindless natural processes is: $I_{nat} = -\log_2[1 - (1 - 0.5)^{1/R}]$. It is possible to determine the likelihood that an effect requires control, where the greater the difference between the functional information required for the effect and I_{nat}, the more likely it is that control was required. This would hold true for SETI, archeology, forensic science, and biological life. For example, any evolutionary process must somehow (regardless of the mechanism) search a sequence space to locate areas where physics produces stable, 3-D structures (500-900 possibilities).

To set an upper limit on abiogenesis I_{nat}, Kalinski assumed the entire mass of the earth consisted of amino acids for building 100-chain proteins, with the entire set reorganizing once per year over the 500 million years of pre-biotic activity, for a total of 10^{55} attempts (R). This results in I_{nat} having estimated maximum of 185 bits of functional information that could result from mindless pre-biotic processes. The biomass of the earth, 7×10^{13} Kg [BioWik], is a tiny fraction of the earth's mass, so fewer than 10^{39} amino acids would actually have been available. The I_{nat} maximum would be lower if realistic values for the available components were used and thermodynamics were considered (no enzymes present to catalyze polymerization). It would be higher if one assumed multiple functional possibilities or more trials. Using Kalinski's estimations, the relative comparison of probabilities (controlled/natural) can be calculated from the difference in observed-natural functional information: ratio = $2^{(I-185)}$.

The watermarks that were found [Mad08] embedded in the Venter Institute's synthetic genome for M. genitalium are formed by choosing base pairs that translate to the single-letter amino acid codes spelling: VENTERINSTITVTE, CRAIGVENTER, etc. (Note "U" is not an amino acid abbreviation, so "V" is used). The 20 possibilities for each position in each word (with only 1 correct/functional combination) result in $I(E_x) = -\log_2(20^{60}) = 259$ bits of functional information. It is therefore more than $2^{(259-185)} = 10^{22}$ times more probable that control produced the watermarks than pure physicality, which is known to be true in this case.

The estimated frequency of occurrence of stable, folded functional protein domains is between 10^{-64} to 10^{-77} [Axe04], which corresponds to 213 256 bits of functional information required to code for a stable, folded protein domain. It is at least 10^{19} times more probable for control to produce a folded functional domain than pure physicality. Using proteins RecA and SecY, the functional information of an average 300-amino acid protein is estimated at 700 bits. Control is *"10^{155} times more probable than mindless natural processes to produce the average protein. Again, if natural selection is invoked to explain the origin of proteins, a fitness function will be necessary"* [Kal08p11].

Using the 700 bits per protein with the estimated 382 protein-coding genes that the simplest known life form requires [Gla06] yields an estimated 267,000 bits for the simplest known life. This information indicates it is about $10^{80,000}$ times more likely that control *"could produce the minimal genome than mindless natural processes. Again, if one wishes to explain the origin of the simplest life form by natural selection, a fitness function will be required that is capable of generating 267,000 bits of functional information"* [Kal08p11]. Note that even if factors of 10^7 more trials and 1000 more functional results were added, and the minimum genes were halved, control would still be $10^{40,130}$ more probable than mindless natural processes as the cause of the simplest life. Also, thermodynamics, enzymes, and other factors are not considered.

Appendix F: What Happened to Darwinism?

Junk DNA, non-coding intragenetic DNA regions, were thought to have no role in the production of proteins. Darwinsts thought regarded them as merely junk. However, research has proven that they play vitally important roles. *"For years, more and more research has, in fact, suggested that introns are not junk but influence how genes work... introns do have active roles"* [Ray03]. *"Geneticists have long focused on just the small part of DNA that contains blueprints for proteins. The remainder – in humans, 98 percent of the DNA – was often dismissed as junk. But the discovery of many hidden genes that work through RNA, rather than protein, has overturned that assumption. These RNA-only genes tend to be short and difficult to identify. But some of them play major roles in the health and development of plants and animals. ... Some scientists now suspect that much of what makes one person, and one species, different from the next are variations in the gems hidden within our 'junk' DNA"* [Gib03].

"Just when scientists thought they had DNA almost figured out, they are discovering in chromosomes two vast, but largely hidden, layers of information that affect inheritance, development, and disease" [Gib03]. Genetic biologist Wojciech Makalowski states, *"Now, more and more biologists regard repetitive elements as a genomic treasure... that repetitive elements are not useless junk DNA but rather are important, integral components of eukaryotic genomes"* [Mak03].

"Nonprotein coding RNA (ncRNA) refers to mRNA that is transcribed from DNA but not translated into protein. Rather than being 'junk' DNA (i.e. an evolutionary relic) some nonprotein coding transcripts may in fact play a critical role in regulating gene expression and so organizing the development and maintenance of complex life" [Per05]. *"Scientists are puzzling over a collection of mystery DNA segments that seem to be essential to the survival of virtually all vertebrates. But their function is completely unknown. The segments, dubbed 'ultraconserved elements', lie in the large parts of the genome that do not code for any protein. Their presence adds to growing evidence that the importance of these areas, often dismissed as junk DNA, could be much more fundamental than anyone suspected"* [Pea04].

"If one adds together nucleotides that are individually nonfunctional, one may end up with a sum of nucleotides that are collectively functional. Nucleotides belonging to chromatin are an example. Despite all arguments made in the past in favor of considering

heterochromatin as junk, many people active in the field no longer doubt that it plays functional roles... Nucleotides may individually be junk, and collectively, gold" [Zuc97].

A new RNA regulatory role that relies on RNAs' ability to communicate with one another dramatically increases known functional genetic information [Pol10]. *"The new findings suggest that nature has crafted a clever tale of espionage such that thousands upon thousands of mRNAs and noncoding RNAs, together with a mysterious group of genetic relics known as pseudogenes, take part in undercover reconnaissance of cellular microRNAs, resulting in a new category of genetic elements which, when mutated, can have consequences for cancer and human disease at large"* [Pol10]. *"Because this new function does not depend on the blueprint that RNAs harbor in their protein-encoding nucleotide sequence, the discovery additionally holds true for the thousands of noncoding RNA molecules in the cell... This means that not only have we discovered a new language for mRNA, but we have also translated the previously unknown language of up to 17,000 pseudogenes and at least 10,000 long non-coding (lnc) RNAs. Consequently, we now know the function of an estimated 30,000 new entities, offering a novel dimension by which cellular and tumor biology can be regulated, and effectively doubling the size of the functional genome"* [Pan10].

The trilobite eye in the lowest fossil stratum had advanced optics, including bifocality. It has produced observations like, *"Trilobites had solved a very elegant physical problem and apparently knew about Fermat's principle, Abbe's sine law, Snell's laws of refraction and the optics of birefringent crystals"* [Cla75]. In a description of *"the advantage of good eye design"* (with a note using design *"as a lead-in to the parallels between the optic designs of humans and the remarkably evolved morphology of trilobites"*), *"the rigid trilobite doublet lens had remarkable depth of field (that is, allowed for objects both near and far to remain in relatively good focus) and minimal spherical aberration"* [Gon07].

Physicist Riccardo Levi-Setti observes, *"In fact, this optical doublet is a device so typically associated with human invention that its discovery in trilobites comes as something of a shock. The realization that trilobites developed and used such devices half a billion years ago makes the shock even greater. And a final discovery – that the refracting interface between the two lens elements in a trilobite's eye was designed in accordance with optical constructions worked out by Descartes and Huygens in the mid-seventeenth century – borders on sheer science fiction"* [Lev93p57]. *"The trilobites of the Cambrian already had a highly advanced visual system. In fact, so far as we can tell from the fossil record thus far discovered, trilobite sight was far and away the most advanced in Kingdom Animalia at the base of the Cambrian... the lenses of the eyes of living trilobites were unique, being comprised of inorganic calcite... There is no other known occurrence of calcite eyes in the fossil record"* [FM-trib].

"There are three recognized kinds of trilobite eyes...with the great majority of trilobites bearing holochroal eyes... characterized by close packing of biconvex lenses beneath a single corneal layer that covers all of the lenses. These lenses are generally hexagonal in outline and range in number from one to more than 15,000 per eye" [Geo-web]. Some believe the abathochroal and schizochroal trilobite eye types evolved from holochroal [Tho05], but there is no fossil evidence to indicate that, and all three types, as well as eyeless trilobites, are found in the same strata. *"Rarely, trilobites may have visual surfaces of normal size, but which lack lenses. This confirms that visual surface growth must have been regulated separately from lens emplacement, and is a feature that cannot be accounted for by the existing developmental model"* [Tho05].

All of the lens' attributes require considerable new prescriptive information somehow to be inserted into the genome in order to indirectly manufacture the lenses. Since no precursors of trilobites are evident, it's difficult to speculate how an eye as complex as a trilobite could have arisen naturally, near the very beginning of life. *"The algorithm must be written in some abstract language... No currently existing formal language can tolerate random changes in the symbol sequences"* [Ede66].

Horizontal gene transfer between different organisms, even different species, is a non-Darwinian mechanism for change believed by many, including microbiologist and physicist Carl Woese. *"How could modern biology have gone so badly off track?... it is a simple tale of scientific complacency. Evolutionary biology took its modern form in the early 20th century with the establishment of the genetic basis of inheritance: Mendel's genetics combined with Darwin's theory of evolution by natural selection. Biologists refer to this as the 'modern synthesis,' and it has been the basis for all subsequent developments in molecular biology and genetics... biologists were seduced by their own success into thinking they had found the final truth about all evolution"* [Woe10].

Appendix G: Scientific Falsification and Specific Hypotheses

"The criterion of the scientific status of a theory is its falsifiability, or refutability, or testability" [Pop63].

Shannon Channel Capacity Theorem [Sha48] would require falsification before giving consideration as science any scenario proposing an alphabet with lower symbolic complexity than the current codon alphabet (see Appendix D).

[Abe05] Testable hypotheses on Functional Sequence Complexity

A single incident of nontrivial algorithmic programming success, achieved without selection for fitness at the decision-node programming level, would falsify any of these four hypotheses:

#1 Stochastic ensembles of physical units cannot program algorithmic /cybernetic function.

#2 Dynamically-ordered sequences of individual physical units (physicality patterned by natural law causation) cannot program algorithmic/cybernetic function.

#3 Statistically weighted means (e.g. increased availability of certain units in the polymerization environment) giving rise to patterned (compressible) sequences of units cannot program algorithmic/cybernetic function.

#4 Computationally successful configurable switches cannot be set by chance, necessity, or any combination of the two, even over large periods of time.

[Abe06] Testable hypotheses about cybernetic organization

A single observation to the contrary would falsify either hypotheses:

#1 Self-ordering phenomena cannot generate cybernetic organization.

#2 Randomness cannot generate cybernetic organization.

[Abe09P] Testable hypotheses about Prescriptive Information (PI)

A single observation to the contrary would falsify any of these three hypotheses. A single prediction fulfillment of spontaneous formal self-organization (independent of agent/investigator involvement and experimenter control) is all that would be necessary.

#1 PI cannot be generated from/by the chance and necessity of inanimate physicodynamics.

#2 PI cannot be generated independent of formal choice contingency.

#3 Formal algorithmic optimization, and the conceptual organization that results, cannot be generated independent of PI.

[Abe09C] Testable hypothesis about Cybernetic Cut
Physicodynamics cannot spontaneously traverse The Cybernetic Cut: physicodynamics alone cannot organize itself into formally functional systems requiring algorithmic optimization, computational halting, and circuit integration.
A single exception of non trivial, unaided spontaneous optimization of formal function by truly natural process would falsify this.

[Abe09U] Testable Universal Plausibility Metric

If $f\ {}^{L}\Omega_{A}/\omega < 1$, scenario is operationally falsified (infeasible)
where f = the number of functional objects/events/scenarios that are known to occur out of all ω possible combinations within the theoretical maximum metric ${}^{L}\Omega_{A}$
The level (L) of the metric may be "q" for quantum or "c" for chemical. The astronomical subset(A) of the metric may be "u" for universe, "g" for our galaxy, "s" for our solar system, and "e" for earth.
${}^{C}\Omega_{e}$ is the metric for the maximum chemical reactions on the Earth = 10^{70}
${}^{C}\Omega_{s}$ is the metric for chemical reactions in the solar system = 10^{85}
Since f/ω is the probability of a particular reaction sequence,
if probability < 10^{-70} with Earth resources, scenario is falsified .
The Universal Plausibility Principle is *"independent of any experimental design and data set. No low-probability plausibility assertion should survive peer-review without subjection to the UPP inequality standard of formal falsification... The application of the Universal Plausibility Principle (UPP) precludes the inclusion in scientific literature of wild metaphysical conjectures that conveniently ignore or illegitimately inflate probabilistic resources to beyond the limits of observational science"* [Abe09U].

Appendix H: Philosophical Hindrances to Scientific Truth

Dawkins writes, *"The more statistically improbable a thing is, the less can we believe that it just happened by blind chance... Darwin showed how it is possible for blind physical forces to mimic the effects of conscious design, and, by operating as a cumulative filter of chance variations, to lead eventual*[ly] *to organized and adaptive complexity, to mosquitoes and mammoths, to humans and therefore, indirectly, to books and computers*" [Daw82].

In other words, those espousing purely physical causes for everything, believe that the original chance formation of the Universe led to chance formation of life which led to humans which led to computers, etc. Tracing the propositional logic backwards leads to chance causing the computer that is being used to record the thoughts (also caused ultimately by chance) for this book. Ignoring the fact that chance has no "causative effect," where is the proof of such a belief? The challenge is to show from information science that the prescriptive programming and other complex functional information could arise "by chance." That is the science question that should trump any unsubstantiated speculations.

It would be nice to believe that presuppositions and biases are not involved in science, and that scientists would follow the evidence, wherever it took them. It is healthy and proper to thoroughly examine claims that purport to be scientific because anyone can speculate anything, but that doesn't make it true. What should not be tolerated is refusal to evaluate the scientific merits because of personal philosophical or theological views. The reality of formalism (not determined by chance or necessity) is well-documented in peer-reviewed scientific literature, in such areas as "choice contingency" for generating prescriptive information [Abe09P] and "natural conventions" for coding systems [Bar08B]. It has even been proposed as "natural genetic engineering" [Wit09, Sha10] to account for evolutionary changes. Since non-material formalism differs from physicality, many materialists deny its reality, except for the formal equations of physics to characterize physicality.

For example, biologist Richard Lewontin writes, *"Our willingness to accept scientific claims that are against common sense is the key to an understanding of the real struggle between science and the supernatural. We take the side of science in spite of the patent absurdity of some of its constructs, in spite of its failure to fulfill many of its extravagant promises of health and life, in spite of the tolerance of the scientific community for unsubstantiated just-so stories, because we have a prior commitment, a commitment to materialism. It is not that the methods and institutions of science somehow compel us to accept a material explanation of the phenomenal world, but, on the contrary, that we are forced by our a priori adherence to material causes to create an apparatus of investigation and a set of concepts that produce material explanations, no matter how counter-intuitive, no matter how mystifying to the uninitiated. Moreover, that materialism is absolute, for we cannot allow a Divine Foot in the door"* [Lew97]. It should be noted that "formalism" can be purely natural, with no "Divine" required, which is overlooked by materialists who view chance and necessity as the only possible natural operators on matter. They incorrectly view non-material formalism as philosophy or theology, instead of scientific reality.

Although non-scientific theistic implications can be envisioned, that alone doesn't make a non-materialist view "religious." The US Supreme Court has held that the implications of material alone do not make a religion even though those implications *"coincide or harmonize with the tenets of some or all religions"* [Sup80]. It also ruled *"The Establishment Clause stands at least for the proposition that when government activities touch on the religious sphere, they must be secular in purpose, evenhanded in operation, and neutral in primary impact "* [Sup71]. A recent US Appeals Court ruling rejected the claim that *"Texas Education Agency's ('TEA') neutrality policy constitutes an establishment of religion, in violation of the First Amendment's Establishment Clause. Because we find no evidence to support the conclusion that the principal or primary effect of TEA's policy is one that either advances or inhibits religion, we conclude that the policy does not violate the Establishment Clause. As such, we affirm the decision of the district court"* [APP10]. The National Center for Science Education supported the suit, claiming that the TEA policy was endorsing "Creationism" (which it wasn't, but anything potentially bringing Darwinism into question is thought by many naturalists to be a religious stance). An excellent legal review of court rulings is available [Lus09].

Some may object to considerations of non-material formalism because it may have religious compatibility. This should mean that they should be equally adamant against Darwinism, since that is compatible with Atheism, Secular Humanism, and Naturalism, which are definitely religions, as affirmed by the Supreme Court: *"Among religions in this country which do not teach what would generally be considered a belief in the existence of God are Buddhism, Taoism, Ethical Culture, Secular Humanism, and others"* [Sup61]. It also indicated that *"religious beliefs... are based... upon a faith, to which all else is subordinate or upon which all else is ultimately dependent.... Some believe in a purely personal God, ... others think of religion as a way of life"* [Sup65]. This Supreme Court criterion makes "physicality is all there is" a religious belief since it cannot be proven and all else is subordinate to it. A US Appellate court also affirmed that *"Atheism is religion, and the group... was religious in nature even though it expressly rejects a belief in a supreme being"* [App05]. Because of these court rulings, one must use care when presenting any speculative purely physical naturalistic scenario, since transgression could be an establishment violation (with the same warnings as teaching Creationism would bring). Presenting only verifiable science, avoiding unverified speculations, should be safe.

Atheist and evolutionary biologist David Sloan Wilson admits, *"many scientific theories of the past become weirdly implausible...* [and] *are a greater cause for concern because they do a better job of masquerading as factual reality. Call them stealth religions"* [Wil07].

Sometimes science seems openly hostile to anything perceived as religious, ignoring the religious nature of non-theistic beliefs. Francisco Ayala, a former president of the American Association for the Advancement of Science, has stated, *"It was Darwin's greatest accomplishment to show that the directive organization of living beings can be explained as the result of a natural process, natural selection, without any need to resort to a Creator or other external agent"* [Aya94p323]. Atheist and science historian Will Provine writes, *"As the creationists claim, belief in modern evolution makes atheists of people. One can have a religious view that is compatible with evolution only if the religious view is indistinguishable from atheism"* [Pro99].

Biologist George Wald dismissed anything besides physicalism with, *"I will not believe that philosophically because I do not want to believe in God. Therefore, I choose to believe in that which I know is scientifically impossible: spontaneous generation arising to evolution"* [Wal58]. Atheist Michael Ruse states that evolution is *"a full-fledged alternative to Christianity... Evolution is a religion. This was true of evolution in the beginning, and it is true of evolution still today"* [Rus00].

Often Darwinism becomes a de facto religion. *"Directed by all-powerful selection, chance becomes a sort of providence, which, under the cover of atheism, is not named but which is secretly worshiped... is an unfounded supposition which I believe to be wrong and not in accordance with the facts"* [Gra77p107]. *"Our theory of evolution has become... one which cannot be refuted by any possible observations. Every conceivable observation can be fitted into it... Ideas... have become part of an evolutionary dogma accepted by most of us as part of our training"* [Ehr67]. After a lecture, atheist biologist Richard Dawkins is reported to have *"admitted in a Q&A that followed of being 'guilty' of viewing Darwinism as a kind of religion"* [Maz10p97].

Wolfgang Smith (mathematician/physicist) writes, *"I am convinced, moreover, that Darwinism, in whatever form, is not in fact a scientific theory, but a pseudo-metaphysical hypothesis decked out in scientific garb. In reality the theory derives its support not from empirical data or logical deductions of a scientific kind but from the circumstance that it happens to be the only doctrine of biological origins that can be conceived with the constricted world view to which a majority of scientists no doubt subscribe"* [Smi92]. Yockey writes concerning publication of the Miller-Urey [Mil53] paper that they *"asked Destiny for confirmation of their faith. The editors of the journal Science did not realize that they were publishing religious apologetics"* [Yoc05p186].

Philip Skell, chemist and member of the United States National Academy of Sciences, writes, *"Darwinian evolution – whatever its other virtues – does not provide a fruitful heuristic in experimental biology. This becomes especially clear when we compare it with a heuristic framework such as the atomic model, which opens up structural chemistry and leads to advances in the synthesis of a multitude of new molecules of practical benefit. None of this demonstrates that Darwinism is false. It does, however, mean that the claim that it is the cornerstone of modern experimental biology will be met with quiet skepticism from a growing number of scientists in fields where theories actually do serve as cornerstones for tangible breakthroughs"* [Ske05].

Lynn Margulis asks rhetorically if *"heritable variation mostly does NOT come from gradual accumulation of random mutation, what does generate Darwin's variation upon which his Natural Selection can act? A fine scientific literature on this theme actually exists and grows every day but unfortunately it is scattered, poorly understood and neglected nearly entirely by the money-powerful, the publicity mongers of science and the media. Worse, much of it is not written in English or well-indexed. This literature shows that symbiogenesis, interspecific fusions... are more important than the slow gradual accumulation of mutation or sexual mergers"* [Maz10p281].

"The kind of explanation we come up with must not contradict the laws of physics. Indeed it will make use of the laws of physics, and nothing more than the laws of physics" [Daw96Bp15]. *"The feature of living matter that most demands explanation is that it is almost*

unimaginably complicated in directions that convey a powerful illusion of deliberate design" [Daw01]. *"Natural selection is the blind watchmaker, blind because it does not see ahead, does not plan consequences, has no purpose in view. Yet the living results of natural selection overwhelmingly impress us with the appearance of design"* [Daw96Bcover].

Note the importance of evidence for reaching Darwinian conclusions. *"Important as evidence is, in this article I want to explore the possibility of developing a different kind of argument. I suspect that it may be possible to show that, regardless of evidence, Darwinian natural selection is the only force we know that could, in principle, do the job of explaining the existence of organised and adaptive complexity"* [Daw82]. *"Darwinism is the only known theory that is in principle capable of explaining certain aspects of life... even if there were no actual evidence in favour of the Darwinian theory"* [Daw96Bp287-288]. *"The theory of evolution by cumulative natural selection is the only theory we know of that is in principle capable of explaining the existence of organized complexity. Even if the evidence did not favour it, it would still be the best theory available!"* [Daw96Bp317] In other words, belief does not depend on evidence, so don't try to confuse the issue with facts.

Undirected physical processes as an absolute belief may have unintended impacts on society in addition to promoting unsubstantiated speculations as science. Several studies have been done to attempt to ascertain the source of human ethics [Beg09]. *"All appearances to the contrary, the only watchmaker in nature is the blind forces of physics, albeit deployed in a very special way... Natural selection, the blind, unconscious, automatic process which Darwin discovered, and which we now know is the explanation for the existence and apparently purposeful form of all life, has no purpose at all"* [Daw96Bp5].

"The time has come to take seriously the fact that we humans are modified monkeys... In particular, we must recognize our biological past in trying to understand our interactions with others. We must think again especially about our so-called 'ethical principles.' The question is not whether biology—specifically, our evolution—is connected with ethics, but how. As evolutionists, we see that no justification of the traditional kind is possible. Morality, or more strictly our belief in morality, is merely an adaptation put in place to further our reproductive ends. Hence the basis of ethics does not lie in God's will... In an important sense, ethics as we understand it is an illusion fobbed off on us by our genes to get us to cooperate. It is without external grounding... Ethics is

illusory inasmuch as it persuades us that it has an objective reference. This is the crux of the biological position. Once it is grasped, everything falls into place" [Rus91].

Since Darwinism has no purpose and survival of the fittest is "law," it would logically follow that humans should not be concerned with endangered species, since any such species obviously isn't fit to survive. Besides, new species will evolve that will be fit. Since humans are "just" animals, there is no such thing as "human rights," and hence no such thing as a violation of those rights. Democracy is both inefficient and unnatural, so dictatorships should be the prevailing world order, with the strongest at the top.

Anything that enhances one's ability to survive should be allowed, according to Darwinism, so that laws regulating behaviors are inappropriate, including those prohibiting murder, robbery, and rape (after all, an alpha male should be able to pass on his genes to as many females as possible). Defective humans should be killed before being allowed to reproduce so the species remains genetically pure. Hitler used his belief in Darwinism to justify extermination of those he considered below the superior Arian race [Kei47, Wei04], in line with the full title of Darwin's book "On the Origin of Species by Means of Natural Selection, or the Preservation of Favoured Races in the Struggle for Life."

Thankfully, most people are repulsed by those views. Even Richard Dawkins admits, *"there are two reasons why we need to take Darwinian natural selection seriously. Firstly, it is the most important element in the explanation for our own existence and that of all life. Secondly, natural selection is a good object lesson in how NOT to organize a society. As I have often said before, as a scientist I am a passionate Darwinian. But as a citizen and a human being, I want to construct a society which is about as un-Darwinian as we can make it. I approve of looking after the poor (very un-Darwinian). I approve of universal medical care (very un-Darwinian)"* [Daw08L]. This is in contrast to his statement, *"I want to persuade the reader, not just that the Darwinian world-view happens to be true, but that it is the only known theory that could, in principle, solve the mystery of our existence"* [Daw96Bpxiv].

It is intriguing that one of the strongest, if not the strongest, voices supporting Darwinism really doesn't follow those beliefs to where they lead. When attempting to look up [Web87] "integrity" some years ago, it wasn't found until the meaning of "integer" was examined, where "integrity" was found embedded. This was interesting to a computer science teacher who taught that an integer is an unfractionated whole.

Appendix I: Index of Definitions/Descriptions by page number(s)

Algorithm 5,8,15,43
Alphabet 28,34-37
Amino acid 17-18,106
Artificial life program 58-61
ATP 18,20,23
Base (nucleotide) 18-19,25,33,1
Base (numeric) 7,25
Cambrian explosion 66-67,70
Cell 29,48
Chance 2,8,81
Chirality (enantiomers) 112
Chromosome 18,27-28
Code 11,28,36,49
Codon 19,25
Composomes 22,109
Cybernetic vii,40,46,51-53,89
Data 7
DNA 17-18,49,106-107
Enzyme (catalyst) 18,49
Epigenome 27,29,44
Eukaryotic/prokaryotic cell 29
Exponential notation 1,80
Falsify, Falsified 6,113-120
Feasible/Plausible 5,120
Flagellum 71-73
Formalism 41,80,121-122
Fossil record 66,70
FSC/OSC/RSC 41-42
Functional info 8-11,50,113
Gene 19
Genome 17-18,25-27,31,41,75
Genotype 42
Helicases 23
Horizontal gene trans.70,81,118
Infeasible 5-6,120
Information 7,33,48
Introns 25,61
Irreducible complexity 70,73-74
Junk DNA 61-63,116-117
Logarithm 109
Macroevolution 55-56,61,66
Messenger RNA (mRNA) 18
Metabolism (1st) 109
Microevolution 55
Modern Synthesis (MS) 80
Morphology 56.66
Mutation 57,65-66
Neo-Darwinism (MS) 57,61,80
Nuclear pore complex 19-20
Nucleotide (base) 18,25
Operating System 14,48
Organelle 29
Overlapping genes 26
Peptide/polypeptide 18,106,111
Phenotype 46
Phyla/phylum 66-67
Prescriptive info 8,12,39
Probability 1-6
Probable 3,4
Protein 18-20
Protocol 9-11,47,53
Recombinant DNA 74-75
Ribosome 19,29
RNA 18,107
RNA world 21,107
Science 81
Scientific notation 1-2
Semantics or syntax 39-40
Semiotic(s) vii,11,20,40,50
SETI 10
Software en----gineering 15-16
Shannon info 8,12,33-37,111
Translation & transcription 18
Transfer RNA (tRNA) 19,49
Trilobite 68-69,117
Turing Complete 13.16,59,60

LaVergne, TN USA
08 October 2010
200099LV00007B/1/P